Radar Signal Simulation

Richard L. Mitchell

Mark Resources, Inc.
Marina del Rey, California

Radar Signal Simulation

Artech House

Chapter 12 Contributors:
G.E. Pollon and J.F. Walker

THE ARTECH RADAR LIBRARY

Electronic Scanning Radar Systems (ESRS)
Design Handbook
Peter J. Kahrilas
Consulting Engineer, Raytheon Company

Phased Array Antennas
Dr. Arthur A. Oliner
Professor of Electrical Engineering
Polytechnic Institute of New York
Dr. George H. Knittel
Staff Member, MIT Lincoln Labs

RADARS — in five volumes
I.   Monopulse Radar
II.   The Radar Equation
III. Pulse Compression
IV.   Radar Resolution and Multipath Effects
V.   Radar Clutter
David K. Barton
Consulting Scientist, Raytheon Company

Radar Detection and Tracking Systems
Dr. Shahen A. Hovanessian
Senior Scientist, Hughes Aircraft Company

Significant Phased Array Papers
Dr. Robert C. Hansen
President, R.C. Hansen Inc.

Copyright © 1976
ARTECH HOUSE, INC.

Printed and bound in the United States of America.

Library of Congress Catalog Number: 75-31380

Standard Book Number: 1-58053-130-X

Preface

The primary emphasis in this book is on the simulation of realistic replicas of radar signals that occur at various points in a radar system. It is a subject of great importance because of the direct impact it has on the quality of the simulation results; yet signals are seldom treated with high fidelity in radar system simulation. The basic reason for this neglect is the complexity of the subject. Simple approaches either do not have the required accuracy or would be prohibitively costly to implement. It is the intent of the author to provide a systematic approach to simulating radar signals that recognizes the need for both accuracy and computational efficiency.

With few exceptions, only digital computer techniques are discussed. The radar environment composed of multiple fluctuating targets and nonhomogeneous clutter is too complex for an analog treatment. Considering the convenience, high speed, and low cost of modern digital computers, their use in radar simulations is indeed practical.

Digital techniques require a data structure that has no analog equivalent. For example, it is almost mandatory at some point in a digital simulation to use a rectangular gridded structure to describe the radar environment; without such a structure, most elementary operations would be clumsy to program and inefficient to run. Therefore these data structures receive considerable attention in this book.

Computation time is also of the upmost importance in simulating radar signals. The difference in running time between a poorly implemented simulation and an intelligently organized one can be several orders of magnitude. For this reason the emphasis is on fast algorithms, though they still must be consistent with the accuracy requirements. Fast algorithms are also given for most of the common library functions.

Signal theory and random processes play an important role in this book, and some tutorial material is provided. The reader will find extensive use of the Fourier transform and such operations as convolutions and correlations. They are indispensable in any treatment of signals, as most electrical engineers appreciate. But since specific algorithms are given in this book, it would be possible to skip over those sections dealing with the theoretical development and concentrate on the applications.

Processing signals is often only one of several elements in a radar system simulation. These remaining data processing and control elements are usually straightforward to simulate since one generally just duplicates the data flow that takes place in the actual system. It still may be an extensive undertaking, but it is not one that needs an exhaustive treatment in a book.

The techniques described in this book began appearing in print when the author started teaching a short course in radar simulation sponsored by Technology Service Corporation. The author is grateful to Dr. Peter Swerling for the opportunity to teach the course. The author is also indebted to many of his colleagues. Specific thanks are due Chester R. Stone, whose mathematical insight and understanding of a computer are without peer. Thanks also to Joel F. Walker for providing many stimulating discussions on this subject and for reviewing portions of the manuscript. He and Dr. Gerald E. Pollon have their turn in Chapter 12; their contribution is certainly appreciated. Finally, the reader cannot fail to notice the heavy influence of Dr. August W. Rihaczek in this book. Although he did not anticipate it, his systematic approach to the analysis of high-resolution radar extends directly to the simulation of radar signals. The author is also grateful to Dr. Rihaczek for many helpful discussions.

RICHARD L. MITCHELL
Rolling Hills Estates, California

Table of Contents

Introduction

Chapter 1 Radar Simulation Requirements

Chapter 2 Two Approaches to Radar Simulation

Chapter 3 Radar Environment Models

Chapter 4 Signals, Filters and Noise

Chapter 5 Received Signals and Receiver Responses

Chapter 6 Mapping Procedures

Chapter 7 Sampled Signals

Chapter 7 Sampled Signals

(continued)

Chapter 8 Generating Random Sequences

Chapter 9 Random Number Generators and Other Library Functions

Chapter 10 Ground Based Radar Example

Chapter 11 Airborne Radar Example

Chapter 12 Real-Time Radar Signal Simulation

Introduction

Radar simulation is rapidly gaining in poularity. There are several reasons for this, an important one being that modern radar systems are becoming increasingly more complex and less tractable to straight-forward analyses. Engineers are demanding the last dB of performance out of a radar system, and conventional analytical procedures often do not yield the desired accuracy. Moreover, computers are getting bigger, faster, cheaper, and more convenient to use. However, the primary reason for the popularity of simulations is cost. Modern radars designed to operate in complex environments are too expensive to build just to see if they will perform as hoped. Even if a radar set is available, the cost of operating it can far exceed the costs associated with a computer simulation. This is especially true of airborne or spaceborne radars.

Other factors also account for the popularity of simulations. Convenience is obviously one of them, since results are as close as the nearest computer center. A simulation is almost a necessary prerequisite if the radar is designed to operate in remote areas or inhospitable climates. Flexibility is a key advantage of a simulation. Parameter changes that might require extensive modifications of actual equipment are generally very simple to accomplish in a simulation. In addition, a simulation offers complete control over the experiment, in contrast to the very limited control one has when operating actual equipment.

Simulation is just one of many tools available to the radar engineer to solve various problems. The most basic is conventional analysis. For example, suppose we have a single point target that is characterized by only one parameter — its radar cross section (RCS). Given the various remaining parameters in the radar range equation, we can easily compute the power received from such a target. Other effects can be included in the analysis, such as thermal noise and ground clutter. Suppose we have homogeneous terrain

so that the ground clutter is also characterized by a single value — its backscatter coefficient. To compute the clutter RCS appearing in a resolution cell we only need to know the area on the ground intersected by that cell. So far this is not a very interesting problem, but it is typical of the type of problems that lend themselves to analysis. We can make the analysis more realistic by assuming that the target RCS fluctuates according to some statistical distribution, with possibly another distribution describing the clutter fluctuation. Now we can use the radar range equation to compute the *average* RCS of the target and clutter, and we can compute various signal-to-noise or signal-to-interference ratios. With the use of existing sets of curves we can also predict how certain of the radar system functions will perform.

The above type of analysis is useful in designing a radar because we can obtain parametric solutions. At a glance we can determine what the important parameters are, and a specific solution is easily obtained. But such solutions are actually of very limited use. There are many radar problems that are not analytically tractable. For example, how would one determine the performance of a tracking radar analytically when there are mutliple targets? And suppose the ground clutter is not homogeneous, or that the processing in the receiver is not linear. In some cases we might be able to write complicated integral equations that describe the desired effects, but usually these equations alone will not offer any insight to the radar engineer. Although some computer solution may be necessary, it is not proper for our purposes to call such a solution a simulation.

In one sense it is possible to simulate the radar output directly if we know its statistical properties. We can use conventional analytical techniques, even if we have to solve complicated equations, to describe the signal properties at the radar output and then use a Monte Carlo approach in a computer to generate specific responses. In Chapter 2 we will elaborate on this functional approach to radar simulation.

Instead of deriving an analytical solution that is carried to the radar output, we could attempt to duplicate in a computer the signal flow throughout the radar system. In such a simulation we can fly the targets along realistic trajectories and assign various nonhomogeneous properties to clutter. Unlike conventional analysis, where we must generally make certain simplifying assumptions in order to solve the problem, we can now utilize very complex descriptions of the various effects, since the resultant signals will be combined in the computer solution. With this approach we are able to generate replicas of the actual signals; there are no general restrictions that must be imposed in order to obtain a solution. This is the approach we will concentrate on in this book.

In a signal simulation we will describe a given scattering phenomenon by a set of scatterers instead of by some complicated functional expression. There is virtually no limit to the type of radar problems that can be solved by this approach, nor is there any limit to the amount of detail that we can incorporate. We can always increase the number of scatterers to make the simulation more realistic. However, there are practical limits that are mainly a function of cost. Thus, even though there might be 10^6 or 10^9 scatterers in an actual situation, we could not afford to simulate every one of them individually. In a simulation we must be selective, not only in terms of the radar environment models, but also in terms of the computational procedures that are implemented. It is the principal motivation of this book to get the most out of a simulation on a fixed budget. While just about any method of implementing a simulation will work, we will concentrate on only those methods that make efficient use of a computer. If cost were no consideration, there would not be any need for a book on radar simulation.

In this book we will describe a general approach to simulating radar signals. However, this does not mean that a single simulation program will evolve that can be used in all applications. We will see that there are many factors that influence the simulation, and since it is our objective to minimize the simulation cost, we will be encouraged to tailor the approach to the application. The result is that ultimately we will have many computer programs, each designed to solve a particular problem. But there will be many features that are common to all programs. This is the reason why it is possible to define a general approach to simulating radar signals.

The central core of a radar simulation is the model or set of models that describes the scattering and propagation of signals external to the radar. There are numerous models that describe various effects, some on a microscopic level and some on a larger or macroscopic level. All models can be accommodated in a simulation, although we will soon discover that this is the area where we must be most selective in order to avoid unnecessary cost. In Chapter 3 we describe the general properties of scattering models that are important for simulation. The problem for the radar engineer will be to choose the proper models for a given application. We must achieve a balance between detail in the model and simulation cost. And the choice must be in proper perspective with the simulation objectives.

In order to simulate the signal received by the radar we will superimpose signals scattered from various points in the environment. To do this superimposition we must specify the phase of the component signals. This leads us to the use of phasors to describe the signals and of complex arithmetic to process them in the simulation. In Chapter 4 we give a summary of the classical mathematical operations that are used to describe and process radar signals. Two operations are basic to our use — the Fourier transformation and the convolution. They are fundamental in simulating radar signals, and everyone contemplating a radar simulation should be familiar with their properties.

In Chapter 5 we derive the receiver response to a collection of scatterers and show how it can be simplified for various classes of waveforms. However, the solutions formulated in Chapter 5 are not particularly well suited for implementing on a digital computer. But in Chapter 6 we define rectangular gridded formats that do make efficient use of the internal structure of a digital computer. We also derive various mapping procedures that are economical to implement in a simulation. As a consequence of using the rectangular grids we will have certain sampling errors. Since the size of these errors depends on how much expense we are willing to incur, we devote a considerable discussion in Chapter 6 to how to specify the largest acceptable error. In a digital computer solution all signals must be sampled. In Chapter 7 we derive the properties of sampled signals and show various operations that can be performed on them. We also continue the discussion on sampling errors.

Most radar signal simulations will be based on some type of random process, especially for

clutter. In Chapter 8 we develop various techniques for generating random sequences with specific statistical properties. In Chapter 9 we show how the random numbers can be generated and we give algorithms that will enable us to generate them very quickly on a digital computer. These same algorithms can be extended to the conventional library functions as we also show in Chapter 9.

In Chapter 10 we apply the methods of the previous chapters to the simulation of a specific ground based radar; in Chapter 11, to an airborne radar. In Chapter 12 we discuss the general techniques that are required to simulate radar signals in real time.

In this book we will concentrate on the simulation and processing of video level signals, which will generally include all the various forms of the radar signal throughout the radar system up to the detection stage. There are two reasons why we will not spend much time on the processing of detected signals. First, there is not much room for innovation at this point; about all we can do is mimic the processing algorithms employed in the actual radar. Second, the subject is already covered extensively in the literature.

Chapter 1

Radar Simulation Requirements

The subject of radar simulation covers considerable ground. There are many types of radar systems and many applications. The radar can be on the ground, on board a ship, in the air, or in space. There are several system functions that vary from simply detecting the presence of a target to target recognition. There are several different simulation objectives that range from the testing of an algorithm to the performance evaluation of a complete radar system. Real-time computation may be a requirement. To meet these various demands there are many ways that one can construct a radar simulation. Some are more accurate than others, and some are faster than others. And some methods are costly to implement. In general there is no single simulation program that could be used in all cases. A simulation program that is suitable for one application may be entirely unsuited for another. All of these various factors must be taken into account when a radar simulation is undertaken; otherwise, the finished product will be of little practical use.

Before we spend any time developing techniques used to simulate radar signals, it would be desirable to establish the requirements imposed on the simulation. We begin with the radar system function, probably the most important factor affecting the simulation approach.

1.1/ Radar System Functions

The many applications of radar can generally be grouped according to their function. The principal function of radar is *detection* — to detect the presence of a target of interest, usually in the presence of interfering signals. Some interfering signals give the appearance of targets; if detected, they are known as false alarms. In addition to detecting a target the radar can measure the location of the target coarsely in one or several of the radar coordinates. This is usually considered part of the detection function. Search and surveillance are terms also used to designate the detection function.

By performing repeated measurements on a dynamic target the radar can *track* the target along its trajectory. The measurements can be extrapolated to future positions to estimate points of intercept or impact, and extrapolated backwards to estimate points of firing or launch. Only detected targets are tracked. Some radar processors employ a *verification* step prior to tracking to discriminate between likely targets and false alarms, or to prevent further processing of redundant detections of the same event.

Much information can be extracted from the radar signal. It can be processed to take not only trajectory information, but also data on the target size, shape, orientation, spin rate, precession, etc. This general category is designated as *parameter estimation*.

Not all potential targets are of equal interest. Some are more threatening than others (e.g., is it a weapon or a decoy?) and demand more interrogation. A radar can employ *discrimination* to separate the two groups, and this can be on the basis of the signal characteristics or the target location, motion, or its trajectory. If several groups into which the targets are to be sorted are involved, then this becomes *classification*. The term used for uniquely specifying a target is *identification* or *recognition*.

Finally, radars can be used to map distributed target phenomena such as the ground, weather, chaff, or large rigid targets. Here the radar function is *mapping*. Once we obtain a map we can employ data reduction procedures, either automatically or manually, to locate specific targets or target features, to estimate target parameters, and to discriminate, classify, or identify targets.

Most radar systems integrate several functions. For example, targets in track must first be detected and, usually, verified as well. This is important in a simulation because the various functions *interact*. A simulation of each function separately would not lead to the same conclusions regarding total system performance as an integrated simulation would.

1.2/Factors That Affect Simulation

Numerous interrelated factors have a strong influence on the best procedure to be used for a given radar problem. We have already discussed the radar system functions, perhaps the most important consideration. It is listed first, followed by seven other factors that are given in their approximate order of importance:

1) Radar System Function
2) Simulation Objective
3) Budget
4) Interactive Requirements
5) Speed Requirements
6) Accuracy Requirements
7) Size
8) The Platform

Sometimes one factor will clearly dominate all others, as in the case of real-time simulations. Usually the dominance is not so clear. We will now discuss each factor.

Simulation Objective

The simulation objective relates to the intended use of the simulation results or data. We can list several possibilities — component design, algorithm testing, system function evaluation, training, testing radars, and integrated system performance. Component design and algorithm testing are basically the same in regard to the requirements on simulation. Both are of small scale and the use of hypothetical and simple data is almost always satisfactory. Of course we

can test algorithms against real data, but then it is no longer a simulation. In system function evaluation the simulation requirements are dictated primarily by the particular function being evaluated. For training, we can have two objectives. The first, and simplest, is to train students with realistic, but hypothetical data. The second is to train someone with data that represents a real situation. In testing radars or in generating signals to be input to radars under test, we are faced with a real-time problem. We can use both simple and complex simulation procedures, depending on the sophistication of the test.

Finally, in simulating the performance of an integrated system involving several distinct radar system functions, we have possibly the most complicated requirement in the simulation procedure. The functions will interact with each other. For example, in the simulation of search and track, decisions will be made whether to search for new targets or to track existing detections on the basis of the number of targets presently being tracked. And the radar can change its mode of operation according to the environment it encounters. Different waveforms can be used to counter different threats, or the system parameters can be altered to adapt to certain changing conditions. The simulation will also be dynamic, in the sense that the system performance will be determined by a time-sequence of events.

Budget

The available budget has a strong influence on how we simulate a radar system. On a limited budget it will not be possible to have a complex and detailed implementation in a large-scale simulation; in fact, it may not even be possible to simulate anything on a large scale. In general, more complex descriptions of the radar and its environment will cost more to implement and run than the simpler models.

Cost is related not only to complexity but to coverage and the number of replications. Some radar problems are expensive to simulate even with simple models. For example, in the simulation of false alarm probabilities where the expected probability is extremely low, we would require a very large number of replications (more than the reciprocal of the probability) to make a reliable estimate. In such cases it would be more effective to rely on analytical techniques even though there may be severe limitations on their applicability.

Interactive Requirement

A simulation can interact with a radar operator or actual equipment. Under such a requirement the simulation generally must run in real-time. The operator or equipment will issue commands that are based on the present state of the radar output, and those commands will affect the state of the radar. The equipment could be a radar system where the simulation generates signals from the environment to be input into the receiver, or it could be a navigation or a guidance system. Sometimes it is possible to slow down the real-time clock; in such cases, the simulation would run more slowly than real-time. However, in most cases we will not have that advantage and we will be forced to run in real-time. This is especially true if an operator or navigator is in the loop.

A simulation can be interactive internally, as in the simulation of an integrated system. An example is where the radar must react to the state of the environment. However, this consideration has no bearing on a real-time requirement.

Speed Requirements

Other than the real-time requirement, if there is one, there are usually no demands on speed *per se*, except the way in which speed affects cost. The faster a simulation is able to run, the less costly it will be. Thus, in large-scale simulations, or in simulations involving many replications, we will always be interested in speed. We can usually sacrifice some accuracy to increase the

speed of the simulation, and thus lower the cost. In real-time simulations the speed is fixed. The question then is how much accuracy and system complexity should be implemented.

Accuracy Requirements

There are different sides to the question of accuracy. One has to do with fidelity in representing the real world; this aspect affects the models of the radar and its environment. A second aspect of accuracy concerns the inherent precision of the radar. A third deals with the precision of basic computations within the simulation. And a fourth is the accuracy required by the simulation objective. Usually it is desirable to achieve some balance among these various criteria. For example, it is not necessary to compute the basic library functions to ten significant decimal places if the radar is capable of handling only four. Some cost saving can be achieved by simplifying such basic computations within the simulation. Similarly, the accuracy in the models should be tailored to the simulation objective. Small-scale simulations, such as the testing of an algorithm or the evaluation of the performance of a single radar system function, are generally not expensive to run; consequently, we can incorporate a much more detailed description of the radar and its environment than we could if the simulation were of large scale. In a small-scale simulation we are usually looking for detailed performance information about a specific target or event; for simulations of larger scale, such as that of an integrated system, there will be several targets and no one of them is so uniquely important. We usually do not require the same accuracy on any one segment of an integrated system simulation that we would if we were simulating that segment alone. Accuracy can also be affected by the scale of the area being covered by the radar. An example is reconnaissance, where we are focusing our attention on a small area, even though the actual radar might cover a much larger area. Here we are interested in fine-scale detail. At the other extreme, navigation requires a different type of accuracy. We are no longer interested in any one particular feature, but in the overall impression of a larger area.

Usually there is a direct relationship between the accuracy achievable in a simulation and the complexity of the approach used. A greater amount of detail incorporated into the models of the radar and its environment permits us to treat them with more realism (assuming we have sufficient data on what the actual conditions are). We can also have considerable detail in our approach even if we are not attempting to duplicate actual conditions. For example, we can have hypothetical models of the radar environment that are detailed in order to exercise or test complex algorithms in the simulation.

The quality of the data upon which the models are based is an important factor in determining the complexity of the approach to be used, especially if the objective is to duplicate actual measurement results. We will not be able to obtain any higher quality information from the simulation than is inherent in the original models.

Size

There is a factor, or a set of factors, that determines the labor required to construct a given simulation, whether it will be measured in man-days or man-years. The *size* of the job is a measure of effort. The terms *complexity, scope,* and *scale* can also be used, although each term can have other, slightly different meanings in a simulation.

The term *complexity* generally refers to the amount of detail required of, or incorporated in, a simulation. Complex models of the radar environment are required for the more sophisticated radar system functions. And sophisticated processing algorithms require a more complex or detailed implementation.

Scope is related to the simulation objective, referring to the extent or the variety of problems that can be accommodated by a given simulation procedure or program. Most radar simulation programs will be of narrow scope in that they will address a specific problem.

The term *scale* has two meanings. One is the number of functions being simulated, in the sense that the simulation of a single function is of small scale while the simulation of an integrated system would be of larger scale. The second meaning refers to coverage. If we concentrate on detail in a single target or area, we are simulating on a *fine* scale; if we are interested in gross features at the expense of detail, then we are simulating on a *coarse* scale.

Size can also refer to the area of coverage or the number of runs. Thus we can have a rather simple approach directed towards the simulation of a large area. In one sense the simulation is small, but in another it is large. Size can also refer to the number of program steps, the storage requirements, or the volume of input data.

The Platform

The location of the radar platform has some effect on the simulation. It affects the choice of ground clutter models (as we will see in Chapter 3) and also certain operations within the simulation (Chapter 6). The differences are not major as far as general procedures are concerned. However, in practice the simulation programs will end up being quite different for ground based and airborne radars, because the nature of the problem that requires simulation is usually different for each case. For example, ground clutter is often the basic limitation to airborne radar performance; for ground based radars, it is not usually a severe problem, but rain can be.

1.3/Implementation

Radar simulations were once entirely analog; now, practically all radar simulations are implemented on digital computers. The reasons go beyond the inherent convenience and reliability of digital equipment — the target and clutter environments confronting modern high performance radars are simply too complex to be simulated with analog equipment. Extended targets, nonhomogeneous clutter, non-Gaussian signals, and large dynamic range signals in particular are best simulated digitally. And digital simulations are inherently more accurate.

We cannot rule out analog techniques entirely. Many real-time simulations require the use of some analog equipment. One obvious case is the simulation of real-time RF or IF signals to be input to a radar under test, since the signals themselves must be analog. Displays are essentially analog devices; filtering and related operations are very efficiently implemented with analog equipment. But while we may be required to use analog equipment for some operations, it would be highly desirable to use digital equipment for the remainder of the simulation.

Convenience, flexibility in the program structure, simplicity of programming, and ease of program maintenance are all strong arguments for general-purpose digital computers. And we will find that most radar simulations can be implemented on general-purpose computers, although those computers are relatively slow. We can employ special-purpose digital circuits to greatly increase speed if desired. Some operations such as digital filtering and the Fast Fourier Transform (FFT) are easily implemented with special-purpose equipment.

The general rule for implementing a radar simulation is to do as much as possible with a general-purpose digital computer. If it is not possible to meet the time or cost requirements with general-purpose equipment, then we should try to use special-purpose digital circuits for those operations such as the FFT that are relatively fixed, but time consuming. The use of analog equipment should be avoided unless absolutely necessary. If the simulation output is to be an analog signal, then the conversion from digital to analog should be done as far into the simulation as possible.

The radar simulation techniques developed in this book are structured for digital computers. Considerable emphasis has been placed on developing efficient algorithms so that digital computers, and especially general-purpose digital computers, can be used in most applications of interest.

Chapter 2

Two Approaches to Radar Simulation

There are two basic approaches to radar simulation; one does not utilize the signal phase. Without the phase we are only able to describe the functional nature of the radar; thus, we use the term functional simulation. This type of simulation is of very limited utility, and we will devote only a few pages to describing its properties.

The second basic approach to radar simulation does make use of the signal phase. It is a more complex method than a functional simulation, but it can be used to solve many more problems. In effect, we are able to describe individual signals at various points in the radar system. We use the term video signal simulation to designate this approach since the video signal with phase contains all of the relevant information about the radar environment. We will devote the bulk of this book to the video signal simulation.

2.1/The Functional Simulation

The simplest approach to radar simulation is a functional simulation. It is implemented on the basis of the radar range equation, which describes the received power from a target at a range r as

$$P_R = \frac{P_T G^2 \lambda^2}{(4\pi)^3 r^4} \sigma \qquad (2.1)$$

where P_T is the transmitter power; G, the one-way antenna power gain in the direction of the target; λ, the wavelength; and σ, the radar cross section (RCS) of the target (i.e., the ratio of the power reradiated in the direction of the radar to the incident power density). In Eq. (2.1) we

have not included any system loss. These losses, as described by Nathanson [Ref.1] and Barton [Ref. 2], are due to the following effects:

Radome
Plumbing (including waveguides, duplexers,etc.)
Propagation (scattering and attenuation by atmosphere, rain, chaff, etc.)
Processing (mismatch from optimum receiver)
Fluctuating targets
Beam shape (targets may not occur on-axis)
Range gate straddle (targets may not appear in center of range gate)
MTI processing (and Doppler filter straddle)
Collapsing of resolution cells
Scanning antennas
Multipath interference

Many of these losses are simple to compute. Others, like the multipath interference, may not be. In an elementary sense we can define the computation of Eq. (2.1) and the system losses to be a simulation, although this would be a misuse of the term.

The received target signal competes with interference which takes the form of receiver noise, clutter (scattering from undesired sources), and electronic countermeasures (ECM). The signal-to-noise or signal-to-interference ratio is a measure of how well the target will be detected in the presence of interference. A value of the order of 20 (13 dB) for this ratio will generally be sufficient to detect the target, but this value depends on many factors such as the way the target fluctuates and the way the signal is processed in the receiver. Given the signal-to-interference ratio, it is possible to compute the probability of detection if we make suitable assumptions regarding the target and interference fluctuation statistics. There are numerous references on this subject; just about any book on radar will contain sets of detection curves. An extensive treatment is given by DiFranco and Rubin [Ref. 3], and specific cases are treated by Marcum [Ref. 4] and Swerling [Ref. 5]. See also [Refs. 1, 2, 6, 7 (Ch. 2), and 8]. If no target is present, it is still possible to get a detection (a threshold crossing) because of the fluctuating nature of the interference signal. Such a detection is designated as a false alarm.

The calculations necessary to compute the detection and false-alarm probabilities from the radar range equation, the system losses, and interference do not constitute a simulation in a strict sense, even if they are implemented on a computer. To qualify as a simulation we must attempt to duplicate the output of the detection process, which is either a detection (threshold crossing) or no detection (no threshold crossing). There are actually four situations that can occur, depending on whether a target is known to be present or absent, as we show in Table 2.1. The detected output is easy to simulate in a statistical or Monte Carlo sense. For example, suppose a target is known to be present and the probability of detecting it is P_D. If we generate a random variable u distributed uniformly over (0, 1), then we can define a detection to occur if $u \geq P_D$ or, conversely, if $u < P_D$ then we have failed to detect. If no target is present, then we can apply the same technique on the false-alarm probability P_{fa}. In Fig. 2.1 we sketch the general steps involved in simulating the detected output. With this functional approach the bulk of the effort will be in computing the signal and interference powers from the engagement geometry.

Table 2.1/The Four Possible Situations in Detection

	Detection	No Detection
Target known to be present	Correct Detection	Failure to Detect
Target known to be absent	False Alarm	Correct Action

The real utility of this approach becomes apparent when we are simulating a mission or a battlefield engagement. There can be multiple targets appearing at random instants of time and the RCS of each can be different; moreover, the RCS can be a function of the target aspect. The interfering signals can also be dynamic, especially in terms of the scanning antenna pattern. The output of such a simulation might be a detection report which would list the time and location of the detections and whether a target was actually present. A list of the failures to detect would also be important.

The simulation described above is simple to implement. Only the functional nature of the radar is utilized in the simulation (hence, the term functional simulation). The details involved in the waveform and the signal processor are not explicitly treated; they appear only as certain system losses. But because of its simplicity, this functional approach is attractive for large-scale applications, especially if the radar is a small part of the total system, such as a battlefield engagement, fire control, or interceptor guidance. It is also attractive for some real-time applications where time is at a premium, such as in a training simulator displaying dynamic target scenes.

The functional simulation is also limited by its simplicity. Since the details of the waveform are not implemented, it is not possible to simulate specific signals at various points in the system. Consequently, linear processing is almost always assumed. The functional simulation is basically a description of the *average* power of the signal components — the targets, thermal noise, clutter, and ECM. The radar range equation determines how these components are scaled. To be able to utilize one of the standard detection cases, we must be able to describe the statistics of the *output* signal in terms of those of one of the standard cases. This is often difficult to do in a complex radar environment (e.g., suppose the interfering signal is a mixture of Gaussian and log-normal noise). It is not just a matter of accuracy. In many cases it is not even possible (at least not without considerable effort) to use the functional approach. Two examples are nonlinear receivers and adaptive processors. More-

Figure 2.1/Functional Approach to Radar Simulation

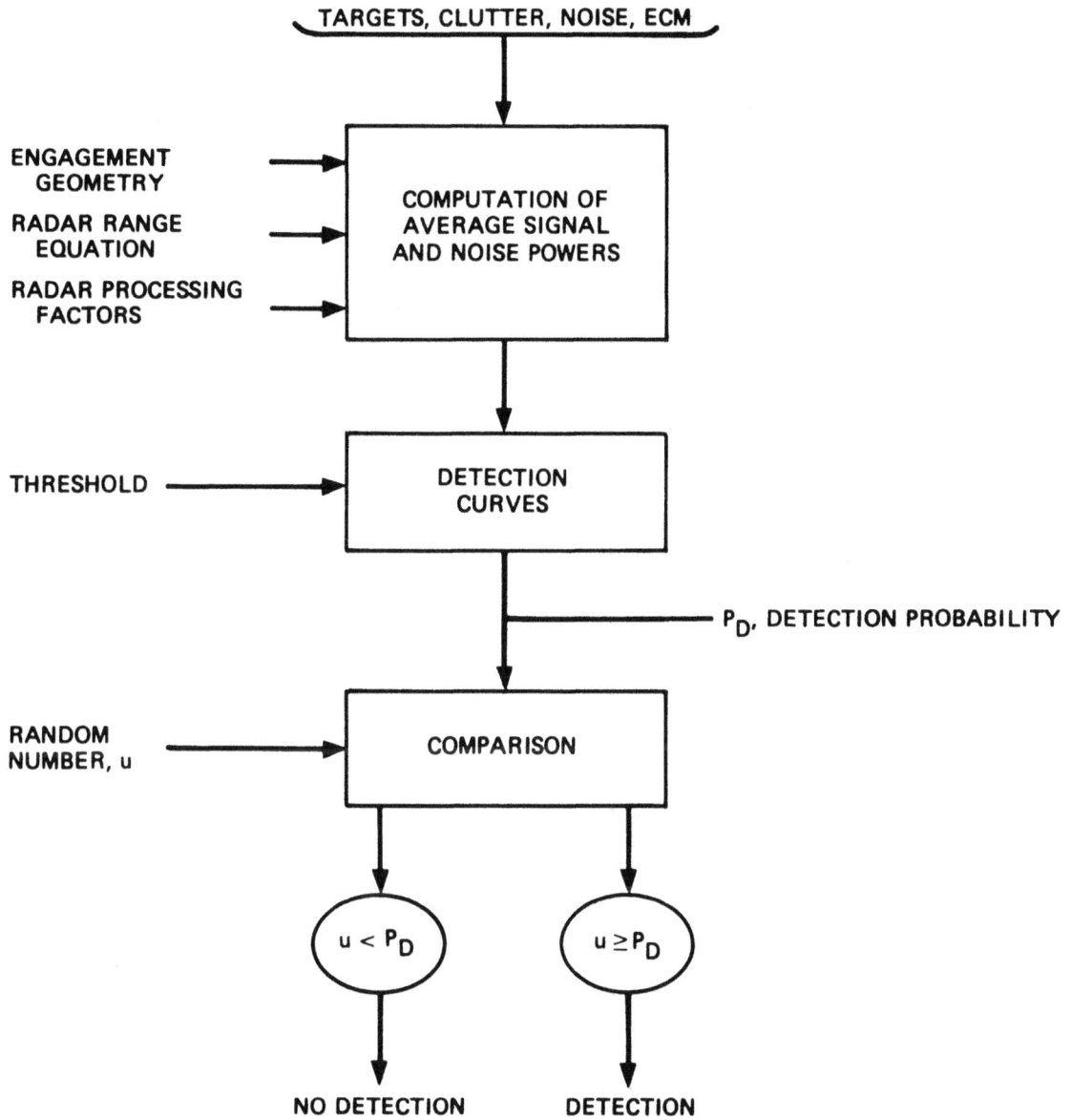

TARGETS, CLUTTER, NOISE, ECM

ENGAGEMENT
GEOMETRY

RADAR RANGE
EQUATION

RADAR PROCESSING
FACTORS

COMPUTATION OF
AVERAGE SIGNAL
AND NOISE POWERS

THRESHOLD

DETECTION
CURVES

P_D, DETECTION PROBABILITY

RANDOM
NUMBER, u

COMPARISON

$u < P_D$

$u \geq P_D$

NO DETECTION

DETECTION

(IF NO TARGET IS PRESENT, THE FALSE ALARM PROBABILITY P_{fa} REPLACES P_D)

over, only point targets (targets that are small compared to a resolution cell) can be simulated. In the next section we will show that it takes a detailed description of the video signal, including the phase, to be able to accurately simulate most of the complex radar systems and environments of interest. In so doing we will pay the price of added complexity in the simulation.

2.2/ The Video Signal Simulation

A video signal simulation is defined basically as an attempt to create a realistic replica of the video signal, consisting of both amplitude and phase, as it is transmitted, propagated through space, reflected from scatterers, and processed in the receiver. We will be able to simulate the actual signal flow in a radar system, although we can utilize linear superposition wherever possible to combine and rearrange elements and thereby eliminate certain computations. A video signal simulation can be implemented with a high degree of accuracy, provided the basic target and radar environment models are good enough. Or in other words, the validity of a video signal simulation is basically limited by the models of the targets and the radar environment.

By simulating the received signal from one scatterer that includes both amplitude and phase, we can simulate the composite signal from several scatterers by superimposing the individual signals. This will automatically simulate the phasor interference among several scatterers. It is also the only effective way to simulate the complex scattering environments that one encounters in dealing with some of the more

sophisticated problems in a high-resolution radar. In most of these cases we are interested not only in the phasor interference among the scatterers, but also in the interaction of the waveform and the antenna with each other and with the environment.

In a video signal simulation we will be dealing with voltage, not power relationships. So instead of describing the reflection properties of a target just by its radar cross section σ, we must also include a phase term that accounts for the phase shift as the signal is reflected. If this phase is ϕ, we can define a complex quantity

$$\gamma = \sqrt{\sigma}\, e^{j\phi} \qquad (2.2)$$

that describes the amplitude and phase reflection properties of the target. For the lack of a better term, we will call γ the *complex reflection coefficient* of the target.* Note that the power associated with γ is the target RCS as

$$\sigma = |\gamma|^2 \qquad (2.3)$$

If we were dealing with just a single point target, then the phase of γ loses its significance. However, if the target consists of several point scatterers, or if there is clutter, then we need to retain the phase to describe the phasor interference among the scatterers.

Before we write the voltage relationship for the received signal, let us rewrite Eq. (2.1) as a function of time to emphasize the fact that the radar range equation is an expression of instantaneous power as

$$P_R(t) = P_T(t-\tau)\, \frac{G^2\lambda^2}{(4\pi)^3 r^4}\, \sigma \qquad (2.4)$$

where τ is the round-trip delay for the target. Now let us transmit an RF signal designated by $\psi_T(t)$. In anticipation of Chapter 4, we use the

---

*No specific name appears to have been given to γ in the radar literature, even though the right side of Eq. (2.2) does appear in print.

complex signal notation so that the power in this complex signal is given by

$$P_T(t) = |\psi_T(t)|^2 \qquad (2.5)$$

The received signal from a stationary point target (before any processing takes place in the receiver) will be a delayed version of $\psi_T(t)$, scaled in amplitude by factors similar to those in Eq. (2.4) as

$$\psi_R(t) = \psi_T(t-\tau) \left[\frac{G^2\lambda^2}{(4\pi)^3 r^4} \right]^{1/2} \gamma \qquad (2.6)$$

Note that by the use of Eq. (2.3) and (2.5), the power associated with $\psi_R(t)$ is identical to Eq. (2.4).

Eq. (2.6) is valid only for a stationary target (stationary with respect to the radar). In addition, it is strictly valid only if the antenna is not scanning across the target. In general, targets will be moving and possibly fluctuating in amplitude and/or phase, and antennas will be scanning, so τ, G, and γ will be functions of time. In this general case we can write Eq. (2.6) as

$$\psi_R(t) = \psi_T[t-\tau(t)] \left[\frac{\lambda^2}{(4\pi)^3 r^4} \right]^{1/2} G(t)\,\gamma(t) \qquad (2.7)$$

This expression is exact (in general, r is also a function of time, but the result is an amplitude modulation that is completely negligible compared to the other effects). Given suitable expressions for $\tau(t)$, G(t), and $\gamma(t)$, we can implement Eq. (2.7) in a simulation. But this is practical only if the number of scatterers is few. The reason is that it is tedious to evaluate $\tau(t)$ for all value of t of interest and then to evaluate $\psi_T[t-\tau(t)]$. In effect, we would be tracing many rays at various instants of time. The problem is further compounded if there are mutliple scatterers. For example, suppose we had 1000 scatterers and we needed to evaluate $\tau(t)$ at 1000 instants of time. This would be a total of 10^6 rays that would have to be computed. Such a computation might take an hour on a high-speed digital computer, and it would simulate the re-

ceived signal for just a single transmission. Even if an analog computer were used, we would require a separate signal generator for each scatterer. Clearly we will have to simplify Eq. (2.7) if the simulation is to be at all practical.

It is possible to simplify Eq. (2.7) if $\tau(t)$, G(t), and $\gamma(t)$ are slowly varying functions of time. Fortunately this will be the situation in most cases of interest. The simplifications, based on the so-called constant-Doppler theory, will allow us to simulate complex environments in a time that could be several orders of magnitude less than if we implemented Eq. (2.7) directly. We will discuss this procedure next.

2.3/ Simplifications Based on Constant-Doppler Theory

In Chapter 5 we will derive the signal received from a target that is moving slowly enough with respect to the radar that it may be assumed to be travelling at a constant range rate over some time interval of interest. In addition we will assume that G(t) and $\gamma(t)$ are constant over this same time interval. With these assumptions the exact expression in Eq. (2.7) reduces essentially to a delayed and Doppler-shifted replica of the transmitted signal as

$$\psi_R(t) = \psi_T(t-\tau)\, e^{j2\pi\nu t} \cdot \left[\frac{\lambda^2}{(4\pi)^3 r^4} \right]^{1/2} \cdot G \cdot \gamma \qquad (2.8)$$

where τ and ν are the delay and Doppler coefficients (constants) of the target. These quantities can be written in terms of the range r and range rate $\dot{r}$ of the target as

$$\tau = 2r/c \qquad\qquad (2.9)$$

$$\nu = -2\dot{r}/\lambda \qquad\qquad (2.10)$$

where c is the propagation velocity and λ is the wavelength. The target is now completely described by only three parameters: τ, ν, and γ (at least over some time interval of interest). For multiple targets, Eq. (2.8) leads to mapping procedures that will enable the signal to be evaluated at regular intervals. Efficient computational algorithms can then be implemented as we discuss in Chapter 6.

Although Eq. (2.8) is an expression for the RF signal, it would be an equally valid expression for the video signal since we could factor the RF center frequency out of both $\psi_T(t)$ and $\psi_R(t)$. In effect, all of the information will be contained in the quadrature components of the video signal. (Such a signal is often designated as the bipolar video or complex video signal.) This is the reason for designating such a procedure as a video signal simulation. In fact, we can even implement Eq. (2.7) on the video signal by simply translating the center frequency down to dc (we can do this if we use the complex signal notation discussed in Chapter 4).

After we obtain the received signal from all scatterers (it is just the superposition of the individual signals), we can simulate the various signal processing steps that take place in the receiver. Usually we will simulate these steps in the same order as in the receiver, although this is not required for a linear receiver. In a high-resolution radar these steps are likely to include both pulse compression and Doppler filtering. We can also simulate nonlinear operations, if necessary, such as clipping and analog-to-digital conversion. Detection, post-detection processing, tracking, and parameter estimation are functions that are also easily implemented.

2.4/When the Simplifications are Valid

To determine the conditions under which the constant-Doppler theory is valid, we will derive inequalities in terms of system parameters. First of all is the assumption of constant delay τ. The resolution of the radar in range is given approximately by

$$\Delta r = c/2B \qquad\qquad (2.11)$$

where c is the propagation velocity and B is the bandwidth. Now if a target is moving with a range rate $\dot{r}$, the target will move a distance $\dot{r}T$ in a time T, which we will use as the time interval over which we wish to use Eq. (2.8). If this distance is small compared to Δr, or if the magnitude of $\dot{r}$ satisfies

$$\dot{r} \ll \frac{c}{2TB} \qquad\qquad (2.12)$$

then we can treat τ as a constant in Eq. (2.8). In other words, the fact that a target is moving only slightly through a range gate means that we can neglect the effect of the motion on the received signal amplitude.

Second is the assumption of constant ν. The resolution of the radar in Doppler is given approximately by

$$\Delta\nu = 1/T \qquad\qquad (2.13)$$

where T is the processing time. Let us assume that this T is the same one as above. If a target is accelerating with a range acceleration $\ddot{r}$, the change in Doppler over the time interval T will be $2\ddot{r}T/\lambda$, which we can derive from Eq. (2.10) by assuming that $\dot{r}$ is a function of time. If this Doppler change is smaller than $\Delta\nu$ in Eq. (2.13), or if the magnitude of $\ddot{r}$ satisfies

$$\ddot{r} \ll \frac{\lambda}{2T^2} \qquad\qquad (2.14)$$

then we can treat ν as a constant in Eq. (2.8). In practice Eq. (2.14) will usually be more restrictive than Eq. (2.12).

Third is the assumption of constant gain G in the direction of the target for a scanning antenna. Let us designate the resolution of the antenna in the plane of the scan as $\Delta\phi$. If the antenna is scanning at an angular rate of $\dot{\phi}$, then it will scan through an angle $\dot{\phi}T$ in time T. If this angle is small compared to $\Delta\phi$, or if the magnitude of $\dot{\phi}$ satisfies

$$\dot{\phi} \ll \frac{\Delta\phi}{T} \qquad\qquad (2.15)$$

then we can treat G as a constant in Eq. (2.8).

To use the inequalities in Eqs. (2.12), (2.14), and (2.15), a factor of 4 will usually be sufficient to ensure that the conditions are satisfied.

Last of all is the assumption of constant γ. Fluctuations in the target reflection properties are usually due to a rotating target presenting different aspects to the radar. If a single parameter γ is used to describe a large complex target, then γ can fluctuate rapidly with time even if the target is rotating slowly. Given the geometry and target motion (this also applies to clutter), we can easily establish whether the assumption of constant γ is valid or not. In such cases where we have a single parameter γ describing a large complex target, we can usually replace the single scatterer by a collection of scatterers where the γ assigned to each would be either constant or a slowly fluctuating quantity as we discuss in Chapter 3.

References

[1]   Nathanson, F.E.; *Radar Design Principles;* McGraw-Hill, New York, 1969.

[2]   Barton, D.K.; *Radar Systems Analysis;* Prentice-Hall, Englewood Cliffs, New Jersey, 1964.

[3]   DiFranco, J.V., and W.L. Rubin; *Radar Detection;* Prentice-Hall, Englewood Cliffs, New Jersey, 1968.

[4]   Marcum, J.I.; "A Statistical Theory of Target Detection by Pulsed Radar," *IRE Trans. Information Theory,* Vol. IT-6, April 1960, pp. 59-267.

[5]   Swerling, P.; "Probability of Detection for Fluctuating Targets," *IRE Trans. Information Theory,* Vol. IT-6, April 1960, pp. 269-308.

[6]   Skolnik, M.I.; *Introduction to Radar Systems;* McGraw-Hill, New York, 1962.

[7]   Skolnik, M.I. (ed.); *Radar Handbook;* McGraw-Hill, New York, 1970.

[8]   Berkowitz, R.S. (ed.); *Modern Radar;* John Wiley & Sons, New York, 1965.

Chapter 3

Radar
Environment
Models

The most important part of a radar system is the mathematical modeling that describes the environment. No matter how well the remainder of the simulation is executed, the overall success will depend primarily on the choice of environmental models. The problem here is that modeling radar phenomena is more of an art than a science. Numerous models have been devised, some good and some not so good, and each one applicable in only certain situations. The problem is further compounded by the fact that, while there is an abundance of models, there will not always be one ready made for the particular simulation of interest. Then one is forced to modify existing models or to construct them from scratch.

Scattering of the radar signal is the dominant phenomenon in a radar simulation. This includes both targets and clutter, and in certain cases multiple scattering effects. Other environmental phenomena are attenuation, refraction, dispersion, and Faraday rotation. Items in this latter group, while of interest in some simulations, are not difficult to model since they tend to be homogeneous in the earth's atmosphere. They can usually be treated as loss factors in the radar range equation (e.g., so many dB attenuation per unit distance through the attenuating medium) or as a rescaling of the radar geometry (e.g., the 4/3-earth radius used for refraction). However, in cases such as propagation in a nuclear-burst environment, everything can be very nonhomogeneous and all propagation effects will then be as difficult to model as scattering, or more so.

In this chapter we will concentrate on scattering models. However, we will not go into detail on any particular one. This would be an extensive undertaking and considerable literature is available on the subject (for example, see [Refs. 1, 2, 3, and 4]). Instead, we will categorize the models and show how each fits into the simulation structure. The objective is to provide the reader with an understanding of how to choose or construct a model in a given situation, what the alternatives are, and how to evaluate the overall effectiveness of the models.

3.1/Classification of Scattering Models

There are many types of scattering phenomena and many types of models. First of all, there are two distinct classes of scatterers — *targets*, which are objects we desire to detect, track, or interrogate in some manner; and *clutter*, which is everything else that is usually deleterious to radar performance. A given object or feature can be of either category depending on the function of the radar. For example, objects on the ground would appear as clutter to an air defense radar but as targets to a ground mapping radar or a radar altimeter. Rain is clutter to all radars except weather radars. The distinction between targets and clutter is important since it affects the way in which the phenomenon should be modeled. Targets are generally few in number and we can afford to expend much more effort in modeling them individually. On the other hand, the number of clutter sources is too great for such a treatment. Clutter must be modeled collectively.

Resolution is the dominant factor that affects the scattering models. Objects that are small compared to the size of a resolution cell can be treated as *point scatterers*, although usually not as being isotropic. Objects that are large compared to a resolution cell are designated as *extended or distributed scatterers*. Both target and clutter can be of either type.

The resolution cell size is not always the determining factor on whether the scattering models are to be of the point or extended type. For example, suppose we are performing a measurement on a target in one of the radar coordinates. The measurement precision can be much finer than the resolution (factors of 10 to 50 are common). In these cases the measurement performance will be affected by extended features of the target. Thus measurement precision, when applicable, supersedes resolution in the determination of the validity of point scattering models.

Scattering models are also classified according to the extent of their statistical content. Some models are completely statistical, such as the well-known Swerling cases for fluctuating targets. At the other extreme we have purely *deterministic* models, where all properties are completely predictable. Deterministic models can be based on actual data, exact theory, approximate theory, or they can be hypothetical. It is their exact repeatability and ability to predict future behavior that makes them deterministic. Statistical models tend to be very simple, in contrast to deterministic ones which tend to be complex, especially if the object doing the scattering is large or has a complex geometry. Since we usually are trying to extract much more information from the radar signal about targets than we are about clutter, the bulk of deterministic modeling will be confined to targets.

In many cases models will contain both statistical and deterministic features. For example, the location or trajectory of a scatterer could be deterministic while the scattering properties are statistical; the gross scattering properties of an object might be described by a smooth deterministic function and the underlying fine structure by a statistical process. The moments of a statistical process can be, and usually are, deterministic. Models incorporating both statistical and deterministic features will be designated as *hybrid* models.

Scatterers exist independently of the radar. Therefore, the most basic way of describing the scattering models would be to do it in their own coordinate system. This approach results in a *physical space* model. Given such a model, it will be the job of the simulation to transform the coordinates to radar space, and this transformation is often time consuming. For the

simpler targets and many types of clutter, it is possible to transform the coordinates *analytically* to those of the radar. This transformation includes a specific encounter geometry (or a parametric description) and some of the radar parameters such as the antenna patterns and waveform resolution properties. If the result of the analytical transformation is still a mathematical or statistical description of the scattering that takes place, the model will be a *radar space* model. Now no transformation will have to be implemented in the simulation. Point targets can be modeled either way, but extended targets are almost always described in physical space. A stationary target can often be modeled in radar space, but a spinning and precessing one could not. Clutter can also be modeled in either coordinate system, although clutter to an airborne radar is best treated in physical space.

Sometimes a model can contain both physical space and radar space descriptors. For example, the RCS properties could be expressed in physical space while the time fluctuation properties are expressed in radar space; often, this procedure is followed for land clutter. Deterministic features of targets and clutter are usually modeled in physical space, in contrast to statistical features, which are more easily transformed analytically. While radar space models are usually more convenient and less costly to implement, they do not have the flexibility of the physical space models that are unencumbered by many of the radar system parameters and the encounter geometry.

Some complicated simulations are best done in stages. In this manner we can avoid in the later stages most of the computations that occur in the early stages. As an illustrative example, ground clutter may be expressed completely in physical space in the first stage of a ground based radar simulation. As part of the simulation in the first stage, the ground clutter will be transformed to the radar coordinates of range, azimuth, and elevation. Strictly speaking, this transformation is not analytical, although the new description of ground clutter could be used as the input ground clutter *model* in further simulations of that radar. The new model will be a radar space one; it will be less flexible than the original physical space model since it will be completely dependent on the radar site, the antenna patterns, and the waveform resolution properties.

3.2/ Point Targets

A point target is an object that is physically small compared to a resolution cell. For such targets there are three distinct regions of behavior that depend on wavelength. They are the *Rayleigh, resonance,* and *optical* regions, and the choice is dependent on whether the object doing the scattering is, respectively, smaller than, about the same size as, or larger than a wavelength. Targets are almost always in the optical region, but individual particles that make up volumetric clutter, such as pieces of chaff or raindrops, are usually small enough compared to the wavelength so that they are either in the Rayleigh or resonance region. This distinction is important since objects in the optical region will usually have an RCS pattern that is a strong function of the aspect angle. Objects in the Rayleigh region will usually scatter the radar signal in a relatively isotropic manner.

This strong aspect dependence on the target RCS causes the amplitude of the received signal to fluctuate as the target rotates or moves relative to the radar. In some cases the target aspect will be known and one can predict the fluctuation behavior; however, in many situations it will be unknown at a given instant of time. Although the nominal aspect of the target might be known, there may be random perturbations in its flight, as with an aircraft in a turbulent wind. Moreover, the radar frequency might change while the target is being illuminated, inducing changes or fluctuations in the target amplitude, as we will show later in this section.

We see that there is a need for several different types of scattering models, depending on the application. The models differ mainly in their statistical content and how closely they describe real data. In the remainder of this section we will discuss the various approaches that have been used to model point targets. In general, these models will consist of a description of the RCS behavior as a function of wavelength, aspect, and polarization. No other radar system parameters will affect the models of point targets (provided the object remains a point target when the resolution properties are changed; for frequency modulated signals, the carrier or mean frequency will determine the wavelength used).

Statistical Models
Purely statistical target models are characterized by the probability distribution function of RCS or amplitude. There are four general classes that are popular in the radar literature; these classes are related to physical target features as shown in Table 3.1.

Table 3.1/Four Classes of Statistical Target Models

Class	Physical Feature
nonfluctuating	single isotropic scatterer such as a corner reflector or sphere
exponential RCS (Rayleigh amplitude)	rough surface phenomenon or many random scatterers
chi-square (Rice)	large dominant scatterer plus many smaller random scatterers
log-normal	smooth flat surface — specular type scatterer

The simplest target model is a nonfluctuating point target, but it is seldom used either in radar analyses or simulations unless, of course, the target is a corner reflector or a sphere. Any other configuration will give rise to RCS fluctuations that can be described by a probability distribution function. The exponential RCS (or equivalently, the Rayleigh amplitude) fluctuation is probably the most common one used by radar engineers because it is convenient to use and extensive performance curves are available in the literature. However, its use is not always justified. A more universal class was proposed by Swerling [Ref. 5] — the chi-square distribution, which includes the nonfluctuating and exponential distributions as special cases and provides an excellent approximation to the Rice distribution [Ref. 6]. The well-known Swerling cases [Ref. 7] are also special cases of the chi-square distribution; moreover, detection probabilities are easily computed for chi-square [Ref. 8]. Unfortunately, the chi-square distribution is not sufficiently general to cover all cases of interest. The log-normal distribution does cover most of the remaining instances, however, namely those associated with occasional very strong *specular* reflections that are typical of flat plates [Refs. 9 and 10]. Detection probabilities have also been computed for log-normal targets [Ref. 11].

In addition to the amplitude distribution the fluctuation rate must also be specified because the performance of a radar is very sensitive to the nature of target fluctuations. The fluctuation rate depends on several factors, principally the physical target size, the motion relative to the radar, and whether the radar frequency is changing. The two extremes are *fast* or *rapid* fluctuation, which means that there is no statistical correlation among pulses, and *slow* fluctuation, which means that there is complete correlation among all pulses while the target is being illuminated. Pulses or waveform samples can also be partially correlated. Block correlation is a mathematical convenience used to designate the case where samples are completely correlated over some time interval (the block), but independent from block to block. Autocorrelation functions or fluctuation spectra can also be employed to describe the target fluctuation properties. Only the chi-square family of fluctuating targets lends itself naturally to partial correlation [Ref. 5].

Encounter Geometry

In a purely statistical simulation it is sufficient to specify the coordinates of a point target in radar space (range, angle, and Doppler). But since the radar response is sensitive to the location of the target within a resolution cell (the target may or may not be at the center of the beam), it is realistic in a purely statistical simulation of detection performance to perturb the nominal coordinates of the target by a random amount that is determined by the size of the resolution cell.

More sophistication can be incorporated into the target model by treating the geometry deterministically. This approach can be taken not only for the target trajectory or motion, but for the target aspect that is presented to the radar as well. The RCS of the target can thus be described as a function of target aspect, as in Fig. 3.1. But rather than specify the highly detailed exact RCS pattern vs. aspect including all minor lobes, or some deterministic approximation to it, we can describe the aspect dependence by a smooth function representing a local average as has been suggested by Crispin and Maffett [Ref. 12]. For example, the solid line in Fig. 3.1. represents such an average. Local fluctuations could then be treated statistically. This case would be representative of one where we knew the *nominal* target aspect but had no precise information on the actual target behavior about this nominal value. With this approach, many fewer samples will be needed to describe the RCS pattern. But we now require additional information on the statistical RCS fluctuations about the average. Besides the RCS being a function of aspect, it is possible for the RCS fluctuations to be of one distribution at certain aspects and of an entirely different distribution at other aspects. The fluctuation rate can also change with target aspect. As we will see, the fluctuation rate is highly dependent on the target size and motion relative to the radar. Large targets will result in more rapid fluctuations than small targets, given the same relative motion.

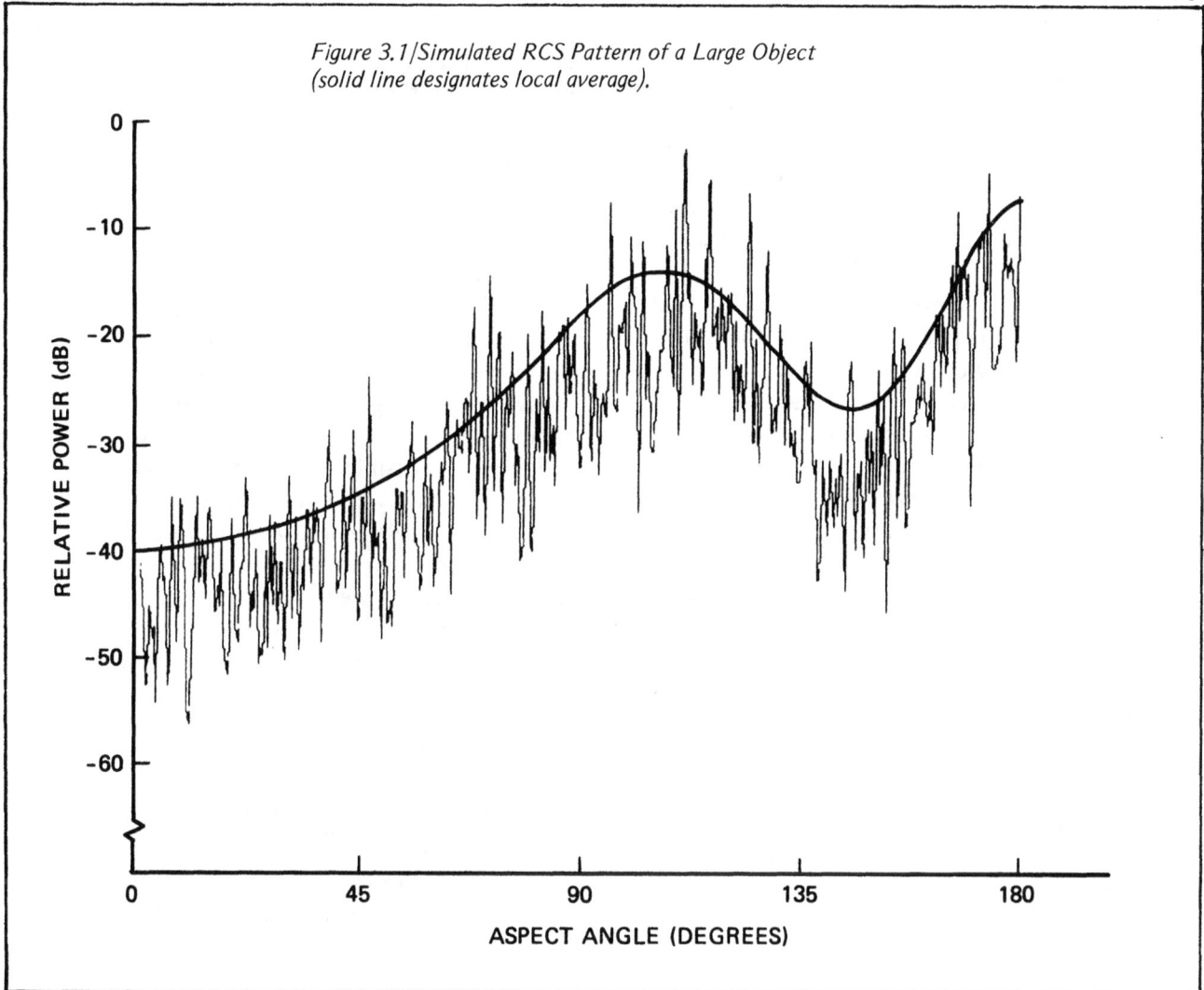

Figure 3.1/Simulated RCS Pattern of a Large Object (solid line designates local average).

Exact Mathematical Modeling

It is extremely difficult to derive exact mathematical expressions for the radar return from complex objects. Such exact solutions, namely ones that would duplicate measurements, demand that the material composition and surface geometry be known with high accuracy. In practice this accuracy is not realistic for most radar targets. Exact solutions are generally based on computing all induced currents within the object and matching boundary conditions. However, this is an extremely tedious undertaking and has been attempted with some degree of success only for the very simplest of shapes —

spheres, cylinders, cones, sphere-capped cones, wedges, etc. [Refs. 1 and 2]. For complex objects there is little hope of ever obtaining exact solutions; even if such solutions were possible, the radar return for most targets of interest would be a rapidly fluctuating function of the aspect geometry as shown in Fig. 3.1. And the engagement geometry will probably never be known accurately enough to predict which minor lobe will be seen at each instant of time. For practical reasons we are thus encouraged to implement models that yield only approximate results unless, of course, we use models that are based on actual measured data.

Approximate Mathematical Models

It is important to be able to construct analytical models of uncooperative targets or of targets that are too large to obtain range measurement data. A practical deterministic approach is to address the nature or the origins of the scattering, namely the *scattering centers*, which occur at or near regions of discontinuity on the surface of the target. If we include the diffraction properties and the phase (relative to the radar) of each scattering center, we will be able to explain many observed phenomena such as the RCS fluctuations in aspect and the polarization dependence of the scattering. This approach is designated as the *geometrical diffraction theory*, or various modifications thereof [Ref. 1 (Ch. 8)]. We will elaborate on the general approach in Section 3.3 when we address extended targets. Modeling methods that do not include the phase of each scattering center will not be able to account for the fine scale behavior of the target RCS.

Models Based on Measured Data

For cooperative targets the most accurate approach to implementing the model in a simulation is to use actual RCS data including the fine structure in the RCS pattern. Two facts limit the use of actual measured data — the difficulty of obtaining such data and the large number of samples required to describe the RCS pattern. Both of these limitations become more serious as the target size increases.

To illustrate how many samples of the RCS pattern are required as we change the target aspect, suppose we have two isotropic point scatterers connected by an imaginary rod of length L and we rotate this rod in a plane that contains the radar line of sight. The reflected signals from the two points will interfere and cause the received signal to fluctuate as we rotate the rod. The most rapid rate will occur when the rod is normal to the line of sight and a rotation of $\Delta\theta = \lambda/2L$ will result in a 2π relative phase change or one complete fluctuation cycle in the RCS pattern. For a more complex target we can define L as an *effective* target length, which might be slightly smaller than the true length. Thus if we sample the RCS pattern in the plane of rotation uniformly at the minimum rate of 2 samples for the shortest cycle, the number of samples required over 360° of rotation is at least

$$N = \frac{2\pi}{\Delta\theta/2} \simeq 25\,\frac{L}{\lambda} \qquad (3.1)$$

If $L/\lambda = 100$, we would need at least $N = 2500$ samples. We could cut this number in half for axisymmetric targets since we would have to sample only over 180° of rotation. A further reduction is possible for long slender axisymmetric targets if we sample uniformly in $\cos\theta$ instead of θ itself, where θ is the aspect angle measured between the target axis and the line of sight. In effect we would be sampling the RCS pattern more finely where the fluctuation rate is highest. Over 180° of rotation the number of samples is now

$$N = \frac{2}{\Delta\theta/2} \simeq 8\,\frac{L}{\lambda} \qquad (3.2)$$

which is about 1/3 less than the number for uniform sampling in θ.

For complex targets that are not axisymmetric we must sample the RCS pattern in 3 dimensions. Thus we would require at least

$$N = \frac{4\pi}{(\Delta\theta/2)^2} \simeq 200\left(\frac{L}{\lambda}\right)^2 \qquad (3.3)$$

samples if we sampled uniformly over 4π steradians. For the above example of $L/\lambda = 100$, N is now 2×10^6 samples. Fewer samples would be needed if we could limit the aspect angle of the target to a certain sector. However, in many situations we would still require an excessive number of samples to describe the actual RCS pattern, especially if we also have to describe the wavelength and polarization dependence of the target RCS.

Use of Raw Video Data

Recordings of the actual video signal can be used instead of a model. Such an approach is realistic and it is the ultimate test of processing algorithms to extract information from the signal. However, such data is too inflexible for most radar simulations. Extrapolation to conditions that are different from the measurement is difficult (such as a change in the PRF) or impossible (a change in the wavelength or engagement geometry). Moreover, recorded video data often contain errors and measurement noise that are not even associated with the radar signal. And it is not possible to acquire good data on uncooperative targets; even when the targets cooperate it is costly to acquire and process the data. Measurement data are more properly used to substantiate existing models, something that can be accomplished with relatively little effort.

Frequency Sensitivity of Targets

So far we have devoted most of our attention to the aspect sensitivity of the radar signal. In some situations the carrier frequency of the radar will change while the target is being illuminated. This tactic is a method of obtaining independent samples of an otherwise slowly fluctuating target (frequency diversity), or it could be the result of a frequency scanning antenna. The target RCS is sensitive to frequency as we can illustrate by the two-point scattering model. Two points separated in range by L will be $4\pi L/\lambda$ radians apart in phase. Since $\lambda = c/f$ where c is the velocity of propagation and f is the frequency, we can equate $4\pi L/\lambda = 4\pi fL/c$. A change in f of $\Delta f = c/2L$ will cause a 2π relative phase change or one complete fluctuation cycle in the RCS. In terms of the wavelength we can write

$$\frac{\Delta f}{f} = \frac{\lambda}{2L} \tag{3.4}$$

If $L/\lambda = 100$, a relative change in frequency of 0.5% will result in one complete fluctuation cycle.

Ordinarily, the fact that frequency can change forces us to add another dimension to the target models. But if the scattering centers of a target are colinear, as on a long slender object, (or if they are coplanar) it is possible to treat a change in frequency as a change in the target aspect. If γ_k represents the complex reflection coefficient of the kth scatterer along the line in Fig. 3.2, the phasor sum of the reflected signals from all the scatterers will be proportional to

$$\gamma_k e^{j4\pi(x_k/\lambda)\cos\theta} \tag{3.5}$$

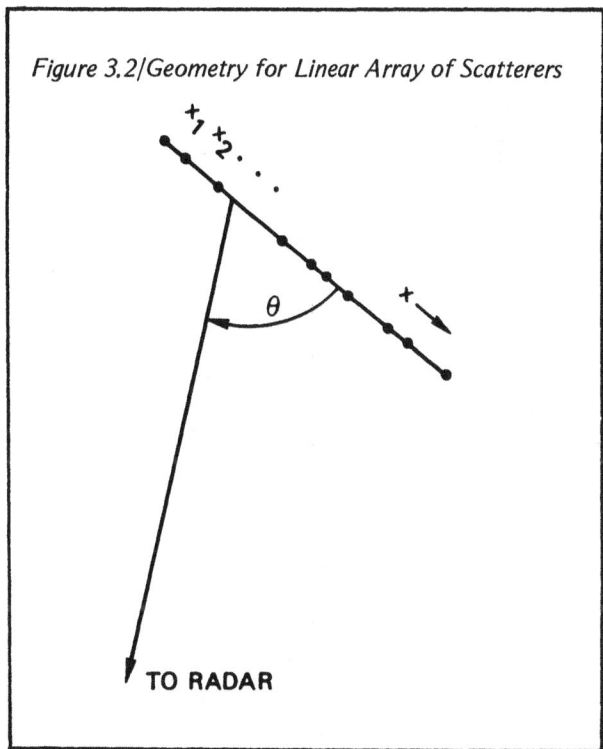

Figure 3.2/Geometry for Linear Array of Scatterers

where x_k is measured along the target axis and θ is measured between the target axis and the line of sight. Since $\lambda = c/f$ we can write the above sum as

$$\sum_k \gamma_k e^{j4\pi(x_k/c)f\cos\theta} \tag{3.6}$$

Note that if f and θ change such that the product f $\cdot$ cos θ is a constant, the composite signal will remain unchanged. The differential relationship is given by

$$\frac{\Delta f}{f} = \tan \theta \; \Delta \theta \qquad (3.7)$$

Given a small change in frequency, Δf, and a target aspect angle, θ, we can solve for $\Delta \theta$ everywhere except in the neighborhood of $\theta = 0$ or π. For these two singular cases, Eq. (3.7) predicts no dependence in frequency since the range extent of the model is now only the size of the assumed point scatterer.

3.3/Extended Targets

The modeling of extended targets is generally more complicated than point targets. For a point target we are concerned with the RCS behavior as a function of only aspect, wavelength, and polarization. But as we increase the target size or decrease the resolution cell size so that the target extends over several resolution cells, we have the additional burden of describing the *spatial* nature of the target. It is no longer possible to describe the RCS behavior of the target as a whole in a functional manner. Instead, we must address the origin of the scattering on the target, namely the *scattering centers* [Refs. 1 (Ch. 8), 3, 13, and 14]. This consideration is also necessary when the measurement precision becomes comparable to, or smaller than, the target dimensions.

Sometimes an extended target can be modeled simply. For example, if the target consists of a few strong scattering centers that can be individually resolved, there will be no interference among them (except through waveform or antenna sidelobes, which do not affect the model). Each scattering center can be modeled as an isolated point target as in the last section. And in many cases only the RCS behavior of the scattering centers will be important since the phase will be removed in the detection process. In such instances, the scattering centers will not have to be located very accurately within the target framework, the precision being determined by the resolution cell size, not the wavelength.

The type of targets that fall into the above category are generally smooth rigid objects where the scattering takes place at edges, corners, or joints. It is straightforward to identify most of the scattering centers; each can be modeled in a deterministic manner, if desired. It is also relatively easy to validate such models against actual data.

On the other hand, purely deterministic approaches do not always work when a target consists of complex or rough surfaces. There will be so many scatterers that there can be several in a resolution cell and the individual scatterers will mutually interfere. The effect is commonly observed in coherent radar images where the response of a homogeneous rough surface area will appear to have strong scattering centers even though there are no dominant scatterers within the area. In Fig. 3.3 we show such an example. The simulated target, in the shape of a cross, consists of a large number of closely-spaced, equal-size scatterers where the phase of each is statistically independent. In the image we see the apparent scattering centers, the locations of which have nothing to do with any physical features on the target. This effect is commonly known as *coherent breakup*.

Because of the phasor interference within a resolution cell, the phase of each scatterer is now important. Moreover, the precision in locating each scatterer is determined by the wavelength. The radar response of such targets will be highly aspect and wavelength sensitive, much more so than that of any one scattering center on the target. At a slightly different aspect a coherent image can take on an entirely different appearance. Scattering centers will fluctuate strongly and appear to move as we change the aspect or wavelength; they can appear in an image at locations where none would be predicted on the basis of the target geometry. That is why such images are usually difficult to interpret.

Figure 3.3/Simulated Coherent Image of Diffuse Target (the target is in the shape of a cross with the radar resolution cell equal to 1/4 of the width of the arm; excerpted from Ref. 13)

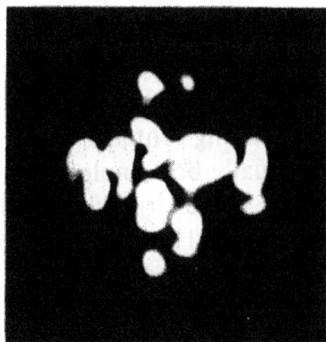

Reprinted by permission of the IEEE.

When the scatterers on an extended target cannot be individually resolved, there are four difficulties in using purely deterministic models to predict the received signal. First, we must specify a *phasor* to describe the scattering properties of each scatterer when, even at best, only a scalar quantity will be known. Second, we must locate each scatterer to a fraction of a wavelength, which is not feasible in practice for large complicated targets. Third, very high accuracy requirements that can barely be achieved in well controlled experiments are placed on the engagement geometry. And fourth, it is not usually possible to validate such deterministic models against actual data. Therefore, there is a strong motivation to use some kind of a statistical approach to treat the phasor interference that takes place within a resolution cell.

Let us assume that the scatterers within a resolution cell are specified by a random complex reflection coefficient with the phase uniformly distributed over 2π rad. This specification is equivalent to assuming that the location of each scatterer is uncertain to within a half-wavelength. If the scatterers are independent the model is reduced to an expression of the *average* RCS for each scatterer with some indication of the probability distribution function for the amplitude. The average RCS for each scatterer can also be a function of the aspect, wavelength, and polarization; this function will generally be smooth, as discussed in the last section.

Since, by definition, it is not possible to resolve the scatterers in a resolution cell, we will lose little generality if we combine the scatterers appearing in a given cell into one new scatterer. Strictly speaking, we should add the individual phasors to create a new phasor. The reason is that the probability distribution of the new phasor will be, in general, different from that associated with the components (the scatterers could represent different scattering mechanisms). But there are two limiting cases of interest where the distribution function of the sum is tractable. The first is the many-scatterer mechanism where there are numerous independent scatterers in a resolution cell with no individual one dominating, leading to Rayleigh-amplitude statistics; it is a common phenomenon for diffuse-type scatterers. The other case is where one of the scatterers in a resolution cell dominates all others. Here the statistics of the composite are determined primarily by the dominant component; examples include specular-type scatterers that often lead to chi-square or log-normal amplitude statistics. For either case, the average RCS of the cell taken as a whole is, or is nearly, the sum of the RCS averages of each component taken separately. If neither the many-scatterer mechanism nor the dominant-scatterer theory is valid, then the random phasors must be added separately.

In order to estimate the average RCS of a resolution cell, we will have to determine the respective origin of scattering. There are several types of target features of interest; the simplest is the diffuse or *rough surface* where the surface irregularities are large compared to a half-wavelength. As we have discussed, the scattering will appear to originate at random locations on the surface, and the scattering will tend to be isotropic. The target feature can also be a *smooth surface* where the surface irregularities are small compared to a half-wavelength. Scattering from such surfaces will be highly directive, especially if the surface is flat. The strong returns that appear at certain aspects are designated as *specular* returns and will be confined to a relatively narrow region of target aspect angles; outside this region, the reflected signal will be relatively weak. Curved surfaces that are smooth do not produce strong specular responses unless the curvature is concave and the radar is located at a focal point. Since the probability of occurrence of the latter is low, it is convenient to treat the specular phenomenon as being associated with flat plates.

Corners formed by flat plates that intersect at right angles are also important target features because the reflected signals can be strong, depending on the area of the plates. Dihedral corners formed by two plates intersecting at right angles act as *line sources* and trihedral corners formed by three plates mutually intersecting at right angles act as *point sources.* If the plates do not intersect at right angles the scattering mechanism is more complicated, but it can be modeled statistically as a rapidly varying function of aspect.

For all practical purposes there are just a few basic scattering mechanisms that are of interest in modeling the scattering within a resolution cell. In fact for most simulations we can define four *basic scattering elements* that will, in combination, describe all scattering effects. They are

1) *Rough surface* — a diffuse scattering phenomenon that is nearly isotropic

2) *Point source* — including trihedral corners and spheres where

the scattering is also nearly isotropic

3) *Line source* — an edge, dihedral corner, thin cylinder, or wire where the scattering is directive in a plane that contains the line source but is relatively isotropic in a plane that is perpendicular to the line source

4) *Flat plate* — a directive or specular type of scattering

Note that depolarization can be included in a model of the line source.

It is difficult to make an exact computation of RCS for any scattering center. For example, the geometry will usually be complex — the boundary of a flat plate will be irregular, or the plate may not be flat at all. Thus it is expedient to have simple scattering models for the basic elements, especially for the line source and flat plate. For these two elements, a smooth function for the aspect dependence, as in Fig. 3.4, will be very convenient to implement. The intensity and width of the specular portion of the response can be functions of the length and area of the line source and flat plate, respectively.

Figure 3.4/Smooth Function for Average RCS Response vs. Aspect

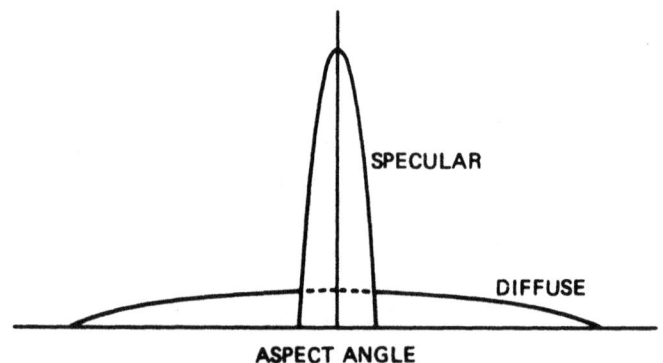

The actual construction of an extended target can be tedious, especially if the target is large. For all but the very simplest of targets (such as in Fig. 3.3) we must describe the target model in the coordinates of the target itself, namely in physical space. A promising approach is to subdivide the target surface into cells that are somewhat smaller than a resolution cell (in Section 6.4 we will discuss the optimum size). Within each cell we will make some determination of what the scattering mechanism is and what basic elements to use. The geometry of the element is important because of the directive properties of line sources and flat plates. Moreover, shadowing must be accounted for since a feature on the surface of the target may not be visible to the radar (creeping and traveling waves are not usually important for extended targets). In order to make the model independent of the encounter geometry, we can construct a list of scattering elements, recording the size, location, and orientation of each in physical coordinates. This data base will be the principal input to a radar simulation, and the first step in the simulation will be to transform the coordinates to those of the radar.

The larger the target is in relation to the resolution cell size, the less important are the details on any particular resolution cell. This situation is shown in the image in Fig. 3.5, which is a simulation of the same target model as in Fig. 3.3, but with a resolution cell that is 1/4 of its former size. The signal features that relate to the geometrical properties of the target are much more important than the cell RCS; less emphasis is needed to specify accurately the RCS of each scatterer in the model. It has also been shown that there are no significant differences in the assumed amplitude distribution function when the resolution is so high [Ref. 13]. Therefore, it is expedient to assume Rayleigh amplitude statistics for all scatterers.

Figure 3.5/Simulated Coherent Image of Diffuse Target (the same target as in Fig. 3.3 but with 1/4 of the resolution cell size; also excerpted from Ref. 13)

Reprinted by permission of the IEEE.

3.4/Land Clutter

Land or ground clutter is a distributed scattering phenomenon, except for *point clutter*, which is associated with man-made structures. The distributed clutter is described by a backscatter coefficient, σ_o, which is an RCS density, and the units are RCS per unit area on the ground (or surface). Given any cell on the ground of area ΔA, the RCS in that cell is $\sigma_o \Delta A$. Point clutter is described by the RCS as if it were a target.

Various approaches can be used to model land clutter. They generally fall into two categories depending on whether the clutter is site-dependent or not. For a ground based radar we are often interested in how well it performs at a given site. Actual terrain features are important in such cases. For shipboard and airborne radars we are generally not interested at all in the performance at one particular site (except when we are testing a radar) since that site may not be representative of the average or worst cases the radar is likely to encounter. We are more interested in a statistical description of ground clutter for such radar.

In modeling site-dependent clutter we must begin with a map of the actual terrain. As we discuss in Section 6.6, the first operation in the simulation will be to transform the terrain elevation to radar coordinates. This information will be used not only to compute the shadow but can also be used to compute the RCS in a given resolution cell on the basis of the local terrain slope. Other information on the terrain map can also be used to compute RCS, such as the type of terrain (rock, dirt, water, etc.) and the type of foliage (trees, brush, grass, etc.). Seasonal and meteorological effects that influence the reflectivity of the terrain can also be taken into account.

Ground clutter is not quite stationary to a ground based radar. The motion of trees, brush, and plants induced by the wind results in a random time fluctuation that is best described in terms of its spectrum. Because of limited measurement data it is expedient to assume that the spectral shape is Gaussian, although Nathanson [Ref. 3] suggests that there is a steady dc component in addition to a fluctuating ac component. The spectral width is roughly proportional to the wind speed with the half-power width being about 3% of the wind speed for wooded terrain [Ref. 3]. In an elaborate ground clutter model we could make the spectral width a function of the foliage type; moreover, we could set the width to zero if there were no foliage.

The global statistics of ground clutter are definitely non-Rayleigh because of shadowing and the presence of occasional strong reflectors. The log-normal and Weibull distributions (see Section 9.1) have both been used extensively. Disregarding the strong reflectors that are actually man-made objects, the statistics within any one cell that is illuminated by the radar will tend to be Rayleigh distributed for the amplitude. Thus, once the average RCS of a cell is known, we have completely specified the clutter model. We can treat the point clutter as a target with a specific aspect dependence, if desired, and it can be modeled deterministically in a site-dependent simulation.

In most analyses of ground based radars, and in some simulations, land clutter is modeled as a homogeneous phenomenon. In these cases σ_0 is usually a function of only the grazing angle; it is a simple matter to analytically transform the clutter to radar coordinates.

The performance of airborne radars, as well as some shipboard, is highly dependent on the nature of nonhomogeneous land clutter. Most of these radars have some type of logic or circuitry that adapts to the varying intensity of ground clutter from one area to the next. It is therefore necessary that a ground clutter model contain nonhomogeneous features if this adaptivity is being simulated. But since it is not important that the precise deterministic nature of the ground clutter be modeled for airborne and shipboard radars, we can construct a hypothetical model on a grid structure that incorporates these spatial nonhomogeneities. This grid must be in ground coordinates, as we will discuss in Section 6.8. In effect we will be constructing a hypothetical ground map of the clutter RCS or RCS density. It is usually satisfactory to specify the *average* RCS density of a given grid square as one member of a random process, and the log-normal distribution function is often used to account for the spatial fluctuations in RCS [Refs. 3 and 15]. We can also make the mean value of the spatial process a function of the grazing angle. In fact, we implement a ground clutter model in Chapter 11 that contains all of these features. Point clutter can be simulated by randomly placing such points on the ground according to some model. Boundaries between one type of terrain and another, especially coastlines, can be treated in a deterministic, but still hypothetical, manner. The map of ground clutter must be transformed to radar coordinates in a simulation.

For ground mapping radars the ground is no longer clutter, but a collection of various types of extended targets. While we can use hypothetical data to construct such models, it will usually be desirable to base the model on real data. This consideration is important in validating the models. For natural terrain we can proceed along the same lines as in the construction of a site-dependent model for a ground based radar. For the cultural features we can model them as extended targets superimposed on the terrain. Because of the large number of cultural features in and around towns and industrial complexes, it is usually satisfactory to model complex objects by a relatively simple solid geometrical shape. While specific detail about any one cultural feature will be lost, the overall appearance in a large area containing many such features will be maintained. Of course, if specific detail about any one feature is desired, as in a reconnaissance mission, then more detail can be incorporated into the various models.

3.5/Sea Clutter

Sea clutter tends to be homogeneous, in contrast to land clutter. Moreover, it is a statistical phenomenon as far as radar performance is concerned. Except for high resolution systems where the individual wave faces can be resolved, sea clutter tends to be Rayleigh amplitude distributed. Other significant factors that influence sea clutter are the sea state or wind speed, the wind direction, the observation time, the grazing angle, the polarization, and the wavelength [Refs. 3 and 4 (Ch. 26)].

Since shadowing, at least in a deterministic sense, is not important for sea clutter, we will not have to describe the clutter in physical coordinates. But for airborne radars it will usually be preferable to retain the physical space description, as we will discuss in Section 6.8.

A Gaussian-shaped spectrum is a satisfactory model for all practical purposes, since there is not much available data to contradict it. Such a model has three parameters, assuming that the amplitude distribution is Rayleigh — the average RCS, the mean Doppler velocity, and the spectral width. As described by Nathanson [Ref. 3],

the average RCS, or actually the average backscatter coefficient, σ_o, is a function of the sea state, the wind direction, the grazing angle, the polarization, and the wavelength. The mean Doppler velocity is a function of the wind direction, being proportional to the cosine of the angle between the wind (or wave) direction and the line of sight. The spectral width is generally proportional to the sea state (the half-power width is about 25% of the wind speed [Ref. 3]), with possibly some weak dependence on the wind direction and polarization.

The spectral width is dependent on the observation time, with the width increasing somewhat as the observation time increases. The explanation appears to be that sea clutter is nonstationary. At one instant of time the instantaneous spectrum will be relatively narrow, but the mean Doppler will fluctuate with time. Such an effect could also be modeled in a simulation of sea clutter.

In high resolution radars it is possible to resolve the individual wave structure. Sea clutter then appears very *spiky* and some regions behind the waves are shadowed at low grazing angles. In these cases a log-normal amplitude distribution is often observed [Refs. 3 and 4 (Ch. 26)].

3.6/Volumetric Clutter

Rain and chaff are the two common types of volumetric clutter, but snow, hail, and (at frequencies higher than 16 MHz) clouds and fog can also appear as clutter. Volumetric clutter is also a distributed scattering phenomenon and it is described by a *radar reflectivity* η, which is the RCS per unit volume. Given a cell of volume ΔV, the RCS in that cell is $\eta \Delta V$.

Volumetric clutter is usually spatially nonhomogeneous. Heavy rain exists primarily in relatively narrow cells of a few kilometers in size. On the other hand, light rain tends to be more uniformly distributed. For simulation purposes we can usually assume that rain clutter is homogeneous within horizontal layers. In effect the radar reflectivity will be a function of only the wavelength, altitude, and rainfall rate. In order to use a parameter that is independent of the wavelength (at least for raindrops that are in the Rayleigh region), it is common to define η in terms of the *reflectivity factor* Z as

$$\eta = \frac{\pi^5 |K|^2}{\lambda^4} Z \qquad (3.8)$$

where $|K| = |(m^2-1)/(m^2 + 2)|$ and m is the complex index of refraction of the raindrops [Refs. 4 (Ch. 24), 16, and 17]. For all practical purposes, $|K|$ is very nearly unity. Many experiments have been conducted to relate Z to rainfall rate R. One of the earliest results, and the one most commonly used, is by Marshall and Palmer:

$$Z = 200R^{1.6} \qquad (3.9)$$

where Z is in mm^6/m^3 and R is in mm/h. However, many alternate relationships exist [Refs. 3, 16, and 17].

Chaff is usually distributed in small coulds or along narrow corridors. A single half-wave or resonant dipole will have an RCS of

$$\sigma = .17\lambda^2 \qquad (3.10)$$

when viewed from a random angle distributed over 4π steradians. However, chaff tends to align itself normal to the wind, so this must be taken into account when computing the RCS density of a chaff cloud. In addition, there are *bird nesting* effects which reduce the total RCS. Chaff dipoles also depolarize the radar signal when linear polarization is used.

Volumetric clutter can be modeled in physical space, which would then require a transformation to radar space as part of the simulation. However, it will usually be possible to analytically transform volumetric clutter to radar space.

This transformation would include the integration over the antenna beam, which can be greatly simplified by treating only the mainlobe since sidelobe considerations are not very significant. For vertical fan beams and at longer ranges we can collapse the altitude extent of the clutter onto the ground and define an equivalent RCS density for the clutter as

$$\sigma_0 = \int_0^\infty \eta(h)dh \qquad (3.11)$$

where h is the altitude above the ground.

The fluctuation statistics of volumetric clutter are almost always Rayleigh because of the large number of independent particles. In some cases when chaff dipoles are much longer than a wavelength, the statistics may deviate from Rayleigh.

The fluctuation spectrum of volumetric clutter is influenced by several effects [Refs. 3 and 16]; the strongest are due to wind and wind turbulence and are independent of the radar geometry. Other noticable effects are related to the varying Doppler frequency of the wind, or of the falling particles, across the finite antenna beam. In a manner similar to land and sea clutter it is common to describe the fluctuation spectrum by a Gaussian shaped function. The mean Doppler frequency is just the Doppler frequency of the wind that appears in the mainbeam of the antenna. Several factors are involved in the computation of the spectral width as discussed by Nathanson [Ref. 3].

3.7/Other Types of Clutter

Radars designed to detect low-RCS projectiles, such as mortar and artillery shells, are also able to detect a wide variety of objects or effects that are lumped into the general category of *radar angels* [Refs. 4 (Ch. 24) and 18]. This grouping includes birds, insects, and anomalous echos from an apparent clear atmosphere. Sometimes these clutter signals appear very much like the signal from an actual target. In such cases a deterministic description is usually required of the motion of each scatterer, although the RCS behavior can almost always be modeled in a statistical manner. If the clutter appears just as background noise, then we can use the general approach described in the last section of modeling volumetric clutter.

Since the term "targets" is used to designate the category of *desired* objects or features, anything else that has the appearance of a target to the radar is often called clutter. Thus booster fragments are sometimes designated as clutter even though they travel with reentry vehicles. Automobiles can appear as clutter to a low-PRF MTI radar designed to detect aircraft, and aircraft can appear as clutter to other types of radars. To avoid confusion it is probably better to designate such objects as *false targets* instead of clutter. These false targets are, in general, deterministic and should be modeled as such. Usually the number of such objects will be few, and we can expend almost the same kind of effort in modeling them as we do for the desired targets.

3.8/ Jamming and RFI

There are two general categories of jamming. The first is *noise-type jamming,* where random noise is generated within a certain band or bands. This type of interference is simulated easily by treating the jamming signal as additional thermal noise; we can also specify the shape of the power spectral density function. In general, the average noise power is a function of only the jammer range and the weighting or suppression by the receive antenna pattern.

The second type of jamming is *deception jamming,* where the jammer attempts to create deterministic signals that have the appearance of target signals. The signal intercepted can be repeated or the jammer can create its own signal. In either case the approach in the simulation is to generate the jamming signals in a deterministic manner.

All radars are affected by stray signals from other radars, and some radars are affected by man-made interference from automobile ignitions and other electrical devices. This general category of signals is designated as *radio frequency interference* (RFI). The signals are sporadic and can be modeled either in a deterministic or statistical manner. Interference from lightening can also be included as a type of RFI.

3.9/ Other Environmental Effects

At certain frequencies or over certain propagation paths there are other environmental phenomena that can strongly affect radar performance. Atmospheric attenuation and multipath are the two most common effects; we will discuss them further in this section. Others that are significant in some cases are refraction at long ranges and low elevation angles. Faraday rotation of a linearly polarized signal propagating through the ionosphere, and dispersion or degradation of coherence of a wide bandwidth signal propagating through a plasma.

Atmospheric Attenuation

There are several sources of atmospheric attenuation. One source that is always present in the lower atmosphere is molecular absorption, principally from oxygen and water vapor. The total two-way attenuation is generally below 1 dB for frequencies below 10 GHz and elevation angles above 5°, regardless of the range. For higher frequencies and lower elevation angles, molecular absorption can be significant. Curves are available that relate attenuation to the frequency, elevation angle, and range [Ref. 4 (Ch. 2)]. Molecular absorption is usually assumed to be homogeneously distributed in horizontal layers [Ref. 4 (Ch. 24)].

Particulate absorption from rain, fog, and clouds tends to be nonhomogeneous in space [Ref. 4 (Ch. 24)]. The attenuation for clouds and fog is proportional to the liquid water content, which means that the resulting models are rather simple to construct. It is more difficult to construct models for rain, mostly because published data are generally valid only over horizontal paths [Ref. 19] and the physics of rainstorms is not well understood. Spatial models do exist [Ref. 20], but there is no universal agreement on their applicability. However, it seems that for many simulation problems we can treat rain as being distributed uniformly within a layer of about 5 km in height. Attenuation from snow and hail is not significant compared to that from rain.

Multipath

Multipath severely limits the performance of radars designed to detect ships or low-altitude aircraft. The phenomenon is straightforward; it is just the phasor interference between the direct ray and a ray reflected from the surface of the earth. In computing the multipath interference, the following factors must be considered — the path length difference, the reflection coefficient of the surface (both amplitude and phase), and the difference, if any, in the target reflection

properties for the two rays. Blake [Ref. 4 Ch. 2)] describes the geometry for both the flat- and spherical-earth cases, and Barton [Ref. 21] discusses in detail the case where the reflecting surface is rough. Sometimes there are anomalous effects that must be included in a model when the reflection point is in the near field of the radar antenna [Ref. 22]. Because of the relatively few multipath situations in a simulation it is always preferable to compute the phasor interference rather than to use existing curves of performance.

3.10/Choosing Models

One of the first problems that one faces in constructing a radar simulation is the choice of models of the radar environment. To someone not familiar with radar simulations the initial approach would probably be to choose the most sophisticated models on the pretext that they would be more realistic, more general, and that better simulation results would be obtained. However, such an approach would be ill-advised. The simulation would not only be too costly to run for most problems, but it could be very difficult to check and interpret the results. The only practical solution is to tailor the complexity of the models to the simulation requirements — simple models for simple problems and more complex models for more complex problems. Because of this diversity, it will not be possible to construct a single simulation program based upon a single set of models that will be usable for all simulation problems.

The dominant factor that affects the choice of models is the radar system function being simulated. The simplest function is detection, and the requirements on environmental models are not high. Statistical models are satisfactory in practically all cases. On the other hand, without incorporating the necessary details in the models, it is not possible to extract the detailed information about a target and its motion that is required in the simulation of a sophisticated radar system function such as target identification. This goal requires a deterministic description of most target features. In the following list we will discuss the basic requirements of the models for the various radar systems functions:

Detection
In a simulation we have control over the location of a target. The principal question is how often the target can be detected. Targets are the dominant scatterers of interest and a statistical description of both the geometry and scattering properties is usually adequate. In some cases it may be desirable to describe the geometry deterministically, including the RCS vs. target aspect. Point target models are sufficient in practically all cases, although there are some radar system designs that will resolve a target into several scattering centers. Distributed clutter plays a less important role in detection when the target is present than when it is absent, but clutter nonhomogeneities are necessary to simulate signals for radar processors that adapt to or are affected by them. Multipath and atmospheric attenuation can be important effects.

False Alarms
When the desired targets are not present in a simulation, detections can still occur from undesired targets and clutter. Both point and distributed clutter are of interest. Statistical models are usually satisfactory for all types of effects, except in the simulation of ground clutter for a ground based radar at a specific site. Interfering signals such as RFI and jammers are important.

Verification
After a detection is obtained it is not necessary to simulate targets and clutter that do not interfere with the resolution cells in the immediate neighborhood of the detection. Targets are of primary interest. Specific details on the target are now important to distinguish it from a false alarm. Specific information on the target trajectory may not be needed in verification, but the target RCS fluctuation properties can be important. Point target models are usually adequate.

Track
Deterministic models of the target trajectory are required for track. Target motion about its center of gravity and the resulting RCS fluctuations can be modeled either deterministically or statistically. Extended features in the models are needed to describe glint and other effects associated with extended targets. Multipath is an important effect for tracking low-altitude airborne targets.

Parameter Estimation
Only extended targets are of interest when one is attempting to extract information about the target other than its trajectory. And the models must usually incorporate a deterministic description of those features that impact on the parameters being estimated.

Discrimination, Classification, and Identification
In general, these sophisticated radar system functions require a deterministic description of the target motion in addition to most of the target scattering properties. This specification applies not only for the desired targets but also for the false targets. For simple targets with a relatively few scattering centers, all scattering can be modeled in a deterministic manner. For complex targets, especially when there will be several scattering centers in a resolution cell, the hybrid methods discussed in Section 3.3 are best applied. Clutter, with the exception of the false targets, is generally not significant.

Mapping
Specific features that are being mapped in the simulation are designated as targets, and they can be of either the point or extended type. Deterministic models are used for the target geometry and for the gross description of noise-like features common with natural terrain, such as the natural boundaries and the elevation profile.

Fine-scale descriptions of natural terrain are treated in a statistical manner. Clutter is usually not treated as a specific scattering phenomenon, other than as a type of background noise. It is also possible to treat waveform and antenna sidelobe effects as background noise, as it is with the waveform ambiguities common with coherent pulse trains.

The radar system function being simulated is not the only factor that influences the choice of models. All of the remaining factors discussed in Section 2.2 affect the models to some degree, mostly in terms of the complexity, realism, and accuracy required of the models. The general rule is to incorporate just enough detail in the models to exercise or test the features in the radar being simulated. This is especially important in large-scale simulations where the cost is of prime concern.

References

[1] Ruck, G.T. (ed.); *Radar Cross Section Handbook, Vols. 1 and 2*; Plenum Press, New York, 1970.

[2] Crispin, J.W., Jr., and K.M. Siegel (eds.); *Methods of Radar Cross-Section Analysis*; Academic Press, New York, 1968.

[3] Nathanson, F.E.; *Radar DesignPrinciples*; McGraw-Hill, New York, 1969.

[4] Skolnik, M.I. (ed.); *Radar Handbook*; McGraw-Hill, New York, 1970.

[5] Swerling, P.; "Recent Developments in Target Models for Radar Detection Analysis," *AGARD Avionics Technical Symposium Proceedings*; Istanbul, Turkey, May 25-29, 1970.

[6] Scholefield, P.H.R.; "Statistical Aspects of Ideal Radar Targets," *Proc. IEEE, Vol. 55*, pp. 587-590, April 1967.

[7] Swerling, P.; "Probability of Detection for Fluctuating Targets," *IRE Trans. Information Theory, Vol. IT-6*, pp. 269-308, April 1960.

[8] Mitchell, R.L., and J.F. Walker; "Recursive Methods for Computing Detection Probabilities," *IEEE Trans. Aerospace and Electronic Systems, Vol. AES-7*, pp. 671-676, July 1971.

[9] Mitchell, R.L.; "Radar Cross Section Statistics of Randomly Oriented Disks and Rods," *IEEE Trans. Antennas and Propagation, Vol. AP-17*, pp. 370-371, May 1969.

[10] Pollon, G.E.; "Statistical Parameters for Scattering from Randomly Oriented Arrays, Cylinders, and Plates," *IEEE Trans. Antennas and Propagation, Vol. AP-18*, pp. 68-75, January 1970.

[11] Heidbreder, G.R., and R.L. Mitchell; "Detection Probabilities for Log-Normally Distributed Signals," *IEEE Trans. Aerospace and Electronic Systems, Vol. AES-3*, pp. 5-13, January 1967.

[12] Crispin, J.W., Jr., and A.L. Maffett; "The Practical Problem of Radar Cross-Section Analysis," *IEEE Trans. Aerospace and Electronic Systems, Vol. AES-7*, pp. 392-395, March 1971.

[13] Mitchell, R.L.; "Models of Extended Targets and Their Coherent Radar Images," *Proc. IEEE, Vol. 62*, pp. 754-758, June 1974.

[14]   Wright, J.W., and A.H. Haddad; "On the Statistical Modeling of Radar Targets," *U.S. Army Missile Command Report RE-72-19*, November 1972.

[15]   Greenstein, L.J., A.E. Brindley, and R.D. Carlson; "A Comprehensive Ground Clutter Model for Airborne Radars," *IIT Research Institute (Chicago, Illinois) Report to Overland Radar Technology Program*, September 15, 1969.

[16]   Atlas, D.; "Advances in Radar Meteorology;" in Landsberg (ed.): *Advances in Geophysics, Vol. 10*; Academic Press, New York, 1964.

[17]   Smith, P.L., K.R. Hardy, and K.M. Glover; "Application of Radar to Meteorological Operations and Research," *Proc. IEEE, Vol. 62*, pp. 724-745, June 1974.

[18]   Flock, W.L., and J.L. Green; "The Detection and Identification of Birds in Flight, Using Coherent and Noncoherent Radars," *Proc. IEEE, Vol. 62*, pp. 745-753, June 1974.

[19]   Medhurst, R.G.; "Rainfall Attenuation of Centimeter Waves-Comparison of Theory and Measurement," *IEEE Trans. Antennas and Propagation, Vol. AP-13*, pp. 550-563, July 1965.

[20]   Atlas, D., and E. Kessler III; "A Model Atmosphere for Widespread Precipitation," *Aeronautical Eng. Review, Vol. 16*, pp. 69-75, Feburary 1957.

[21]   Barton, D.K.; "Low-Angle Radar Tracking," *Proc. IEEE, Vol. 62*, pp. 687-704, June 1974.

[22]   Pollon, G.E.; unpublished results.

Chapter 4

Signals, Filters and Noise

In this chapter we summarize the fundamental properties of signals, filters, and noise — the basic ingredients in a simulation of a radar system. We also introduce the Fourier transform and complex signal notation. These two mathematical devices are so important to the analysis and simulation of radar systems that little could be done without their use.

The material in this chapter is tutorial, and most derivations are avoided in order to be brief. Most of the notation follows that of Rihaczek [Ref.1], who also gives a good introduction to the subject in his Chapter 2. There are several other useful references on the subject [Refs. 2-8].

4.1/Real Signals

All signals implemented in practice must necessarily be of finite energy. Such a signal can be represented as a function of time s(t) or as a function of frequency S(f). Both representations are related through the *Fourier transform pair*

$$s(t) = \int_{-\infty}^{\infty} S(f) \, e^{j2\pi ft} \, df \qquad (4.1)$$

$$S(f) = \int_{-\infty}^{\infty} s(t) \, e^{-j2\pi ft} \, dt \qquad (4.2)$$

It is customary to denote s(t) as the signal or waveform and S(f) as the frequency spectrum, or simply, the *spectrum*, although both representations can be called the signal.

So far we have not stipulated that either s(t) or S(f) be real. The Fourier transform pair relationship is valid without this assumption. In a real signal representation it is customary to constrain s(t) to be real. With this assumption the complex conjugate of Eq. (4.2) is given by

$$S*(f) = \int_{-\infty}^{\infty} s(t) \, e^{j2\pi ft} \, dt \qquad (4.3)$$

Hence, it follows that

$$S(-f) = S*(f) \qquad (4.4)$$

This is an important property of a real signal. If we take the absolute value as |S(-f)| = |S*(f)| = |S(f)|, we observe that whatever shape the envelope has for positive frequencies, it is exactly reflected for negative frequencies. In other words, one-half of the real signal spectrum uniquely specifies the signal.

Let us divide the integral in Eq. (4.1) into two parts as

$$s(t) = \int_{0}^{\infty} S(f) \, e^{j2\pi ft} \, df + \int_{-\infty}^{0} S(f) \, e^{j2\pi ft} \, df \qquad (4.5)$$

By the use of Eq. (4.4) we can write

$$s(t) = \int_{0}^{\infty} [S(f) \, e^{j2\pi ft} + S*(f) \, e^{-j2\pi ft}] \, df \qquad (4.6)$$

Note that the quantity within brackets is real. In fact, we can rewrite Eq. (4.6) as

$$s(t) = \text{Re} \left\{ \int_{0}^{\infty} 2S(f) \, e^{j2\pi ft} \, df \right\} \qquad (4.7)$$

This is another way of illustrating that we need only one-half of the signal spectrum to uniquely specify a real signal.

The use of real signal notation leads to certain complications in the analysis and simulation of radar and communication systems. The general problem is a proliferation of "cross terms" when signals are processed. The problem can be alleviated by utilizing complex signal notation. Moreover, the complex signal notation is natural for use in radar and communication systems because the real and imaginary components of the complex signal correspond precisely to the in-phase and quadrature channels in a receiver.

4.2/ Complex Signals

Let us define a complex signal $\psi(t)$ in such a way that when we take the real part of it we obtain a real signal as

$$s(t) = \operatorname{Re}\{\psi(t)\} \qquad (4.8)$$

If we compare this with Eq. (4.7) we can define

$$\psi(t) = \int_0^\infty 2S(f)\, e^{j2\pi ft}\, df \qquad (4.9)$$

Let us also define

$$\Psi(f) = \begin{cases} 2S(f), & f > 0 \\ S(f), & f = 0 \\ 0, & f < 0 \end{cases} \qquad (4.10)$$

where, for mathematical consistency, we have not doubled the real spectrum at $f = 0$. Now we can write the Fourier transform pair for complex signals as

$$\psi(t) = \int_{-\infty}^\infty \Psi(f)\, e^{j2\pi ft}\, df \qquad (4.11)$$

$$\Psi(f) = \int_{-\infty}^\infty \psi(t)\, e^{-j2\pi ft}\, dt \qquad (4.12)$$

Note that we can always obtain the complex signal from the real signal by the use of Eqs. (4.9) and (4.10). We double the spectrum for the positive frequencies and set the negative frequencies equal to zero. In this way we also obtain an *analytic signal*, namely one that has only positive frequencies.

It is possible to separate the complex signal $\psi(t)$ into one function that contains only low frequencies and another one containing the mean or *carrier frequency* f_c as

$$\psi(t) = \mu(t)\, e^{j2\pi f_c t} \qquad (4.13)$$

The quantity $\mu(t)$ is complex, hence it is designated as the *complex modulation function*. If we substitute this into Eq. (4.12) we obtain

$$\Psi(f) = M(f - f_c) \qquad (4.14)$$

and hence

$$M(f) = \Psi(f + f_c) \qquad (4.15)$$

which are simple translations of the spectra. The quantity $M(f)$ is also related to $\mu(t)$ by the Fourier transform pair

$$\mu(t) = \int_{-\infty}^\infty M(f)\, e^{j2\pi ft}\, df \qquad (4.16)$$

$$M(f) = \int_{-\infty}^\infty \mu(t)\, e^{-j2\pi ft}\, dt \qquad (4.17)$$

Note that $M(f)$ is now centered near dc and has negative as well as positive frequencies in contrast to $\Psi(f)$, which has only positive frequencies. Thus the complex modulation function is not an analytic signal. In Fig. 4.1 we show how the various signals are related.

4.3/ Narrowband Signals

Radar signals are usually narrowband in the sense that most of the signal energy is concentrated near the carrier frequency. In this case we can write the real signal as

$$s(t) = a(t) \cos\left[2\pi f_c t + \phi(t)\right] \qquad (4.18)$$

where $a(t)$ and $\phi(t)$ are the amplitude and phase modulation functions, respectively, which contain only low frequencies compared to f_c. For narrowband signals (also designated as signals with a low percentage bandwidth) $a(t)$ will trace out the envelope of the rapidly oscillating cosine

Fig. 4.1 Spectral Envelopes for Real and Complex Signals

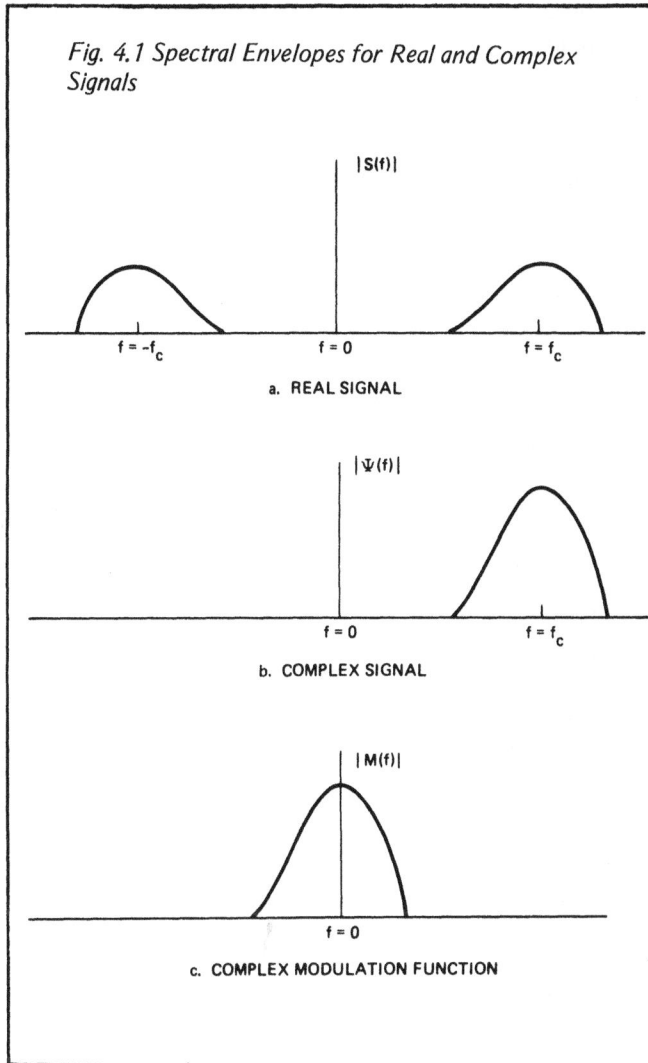

a. REAL SIGNAL

b. COMPLEX SIGNAL

c. COMPLEX MODULATION FUNCTION

$$\psi(t) = a(t)\, e^{j[2\pi f_c t + \phi(t)]} \tag{4.19}$$

This procedure does not always result in a strictly analytic signal, especially if the percentage bandwidth is large. Thus it is expedient to designate it as the *exponential* signal, in contrast to the analytic signal obtained by doubling the spectrum of a real signal for the positive frequencies and setting the spectrum equal to zero for negative frequencies.

By comparing Eq. (4.19) with Eq. (4.13), we can write for narrowband signals

$$\mu(t) = a(t)\, e^{j\phi(t)} \tag{4.20}$$

The reason why this result is limited to narrowband signals is that while we have complete freedom in choosing $a(t)$ and $\phi(t)$ for real signals, their use in Eq. (4.20) will not always produce an analytic signal in Eq. (4.13). Equivalently, while $|\mu(t)|$ is always the envelope or modulus of the complex signal, regardless of the bandwidth, the absolute value of Eq. (4.20), namely $a(t)$, does not represent the envelope of the real signal unless the signal is narrowband. Throughout the remainder of this book, we will assume that signals are narrowband.

4.4/ Filters and Convolution

It is common to define a filter in terms of its transfer function in the frequency domain. This function selectively filters or weights each frequency component of a signal. In complex signal notation, the spectral output of a filter is given by

$$G(f) = \Psi(f)\, H(f) \tag{4.21}$$

function in Eq. (4.18). Thus it is the convention to designate $a(t)$ as the *envelope* of the narrowband signal. For broadband signals it is no longer possible to interpret $a(t)$ as an envelope function, although it is possible to write Eq. (4.18) for any signal.

An alternate method of obtaining the complex signal from a narrowband real signal is to replace the cosine in Eq. (4.18) by an exponential to obtain where $\Psi(f)$ is the spectrum of the input signal and $H(f)$ is the filter transfer function. In general, $H(f)$ is complex.

Given $H(f)$, we can define the impulse response in the time domain as

$$h(t) = \int_{-\infty}^{\infty} H(f)\, e^{j2\pi ft} df \qquad (4.22)$$

and to complete the Fourier transform pair, we write

$$H(f) = \int_{-\infty}^{\infty} h(t)\, e^{-j2\pi ft}\, dt \qquad (4.23)$$

The filter response to a complex signal $\psi(t)$ can also be expressed in the time domain. Let us first take the Fourier transform of Eq. (4.21) as

$$g(t) = \int_{-\infty}^{\infty} \Psi(f)\, H(f)\, e^{j2\pi ft}\, df \qquad (4.24)$$

Now substitute Eqs. (4.12) and (4.23) as

$$g(t) = \iiint_{-\infty}^{\infty} \psi(t_1)\, h(t_2)\, e^{j2\pi f(t-t_1-t_2)} df\, dt_1\, dt_2 \qquad (4.25)$$

To solve this triple integral we will make use of the Dirac delta function $\delta(t)$, which has the property

$$\int_{-\infty}^{\infty} u(t)\, \delta(t-t_0)\, dt = u(t_0) \qquad (4.26)$$

where $u(t)$ is any arbitrary function. If we let $u(t) = e^{-j2\pi ft}$ and $t_0 = 0$ we can write

$$\int_{-\infty}^{\infty} \delta(t)\, e^{-j2\pi ft}\, dt = 1 \qquad (4.27)$$

for all f. Although we can write the inverse Fourier transform as

$$\int_{-\infty}^{\infty} e^{j2\pi ft}\, df = \delta(t) \qquad (4.28)$$

there is some question about its precise mathematical meaning [Ref. 2]. But there is no question about the validity of using Eq. (4.28) in ingegrals such as that in Eq. (4.25). It is therefore convenient to accept Eq. (4.28) as a truism for our purposes [Ref. 1]. Thus Eq. (4.25) reduces to

$$g(t) = \iint_{-\infty}^{\infty} \psi(t_1)\, h(t_2)\, \delta(t-t_1-t_2)\, dt_1\, dt_2$$

$$= \int_{-\infty}^{\infty} \psi(\xi)\, h\,(t-\xi)\, d\xi \qquad (4.29)$$

$$= \int_{-\infty}^{\infty} \psi(t-\xi)\, h\,(\xi)\, d\xi \qquad (4.30)$$

where ξ has been substituted for t_1 in Eq. (4.29) and for t_2 in Eq. (4.30). These expressions are equivalent definitions of the *convolution* of the time-domain signal with the impulse response of the filter. There is also a shorthand notation for the convolution which is written as

$$g(t) = \psi(t) \star h(t) \qquad (4.31)$$

It has the same meaning as Eq. (4.29) or (4.30).

4.5/ Energy

To derive the energy relationships it is first convenient to write the convolution of two arbitrary functions $u(t)$ and $v(t)$ as the Fourier transform of the product of their spectra [Ref. 2] as

$$\int_{-\infty}^{\infty} u(t)\, v(\lambda-t)dt = \int_{-\infty}^{\infty} U(f)V(f)\, e^{j2\pi f\lambda}\, df \qquad (4.32)$$

which we can write immediately from Eqs. (4.24) and (4.29). Now set $\lambda = 0$ and $v(t) = u*(-t)$ to obtain

$$\int_{-\infty}^{\infty} |u(t)|^2 \, dt = \int_{-\infty}^{\infty} |U(f)|^2 \, df \qquad (4.33)$$

where we note that the Fourier transform of $u^*(-t)$ is $U^*(f)$. This result is Plancherel's theorem for Fourier transform pairs.

Now let us define energy for a real signal as

$$E = \int_{-\infty}^{\infty} s^2(t) \, dt \qquad (4.34)$$

With the use of Eq. (4.33) we can write

$$E = \int_{-\infty}^{\infty} s^2(t) \, dt = \int_{-\infty}^{\infty} |S(f)|^2 \, df$$

$$= 2 \int_{0}^{\infty} |S(f)|^2 \, df \qquad (4.35)$$

where Eq. (4.35) is obtained by the use of Eq. (4.4). Since the integral represents energy, the integrand $|S(f)|^2$ is designated as the *energy density spectrum*. For complex signals we will depart somewhat from convention and define the *energy in the complex signal* as

$$E_c = \int_{-\infty}^{\infty} |\psi(t)|^2 \, dt = \int_{-\infty}^{\infty} |\Psi(f)|^2 \, df \qquad (4.36)$$

If we substitute Eq. (4.10) into the second integral of Eq. (4.36) we obtain

$$E_c = \int_{0}^{\infty} |2S(f)|^2 \, df$$

$$= 4 \int_{0}^{\infty} |S(f)|^2 \, df \qquad (4.37)$$

which is twice the value in Eq. (4.35). Therefore

$$E_c = 2E \qquad (4.38)$$

The energy in the complex signal is exactly twice that in the real signal (thus implying that the energy in the imaginary part of the complex signal is identical to the real part). For the complex modulation function, it is easy to show that

$$E_c = \int_{-\infty}^{\infty} |\mu(t)|^2 \, dt = \int_{-\infty}^{\infty} |M(f)|^2 \, df \qquad (4.39)$$

The use of E_c instead of $2E$ avoids the factors of 2 or 1/2 that would appear in all expressions involving energy.

4.6/ Matched Filters and Correlation

A matched filter is commonly derived by maximizing the output signal-to-noise ratio of the filter [Ref. 1]. But for our purposes let us derive one form of the matched filter by maximizing the output response envelope subject to the constraint that the energy in the filter is constant. This approach is equivalent to the former under the assumption of white (broadband) noise.

Let us denote the output signal spectrum by

$$E(f) = \Psi(f) \, H(f) \qquad (4.40)$$

where $H(f)$ is the filter we will optimize. The Fourier transform of Eq. (4.40) is given by

$$\rho(t) = \int_{-\infty}^{\infty} E(f) \, e^{j2\pi ft} \, df \qquad (4.41)$$

$$= \int_{-\infty}^{\infty} \Psi(f) \, H(f) \, e^{j2\pi ft} \, df \qquad (4.42)$$

The response envelope

$$|\rho(t)|^2 = \left| \int_{-\infty}^{\infty} \Psi(f)\, H(f)\, e^{j2\pi ft}\, df \right|^2 \qquad (4.43)$$

is to be maximized subject to the constraint that

$$\int_{-\infty}^{\infty} |H(f)|^2\, df = E_H \quad \text{(a constant)} \qquad (4.44)$$

By the use of the Schwartz inequality for a complex function we can write

$$\left| \int_{-\infty}^{\infty} \Psi(f)\, H(f)\, e^{j2\pi ft}\, df \right|^2 \leqslant \int_{-\infty}^{\infty} |\Psi(f)|^2\, df \int_{-\infty}^{\infty} |H(f)|^2\, df$$
$$(4.45)$$

The first integral on the right is E_c and the second is E_H by virtue of Eq. (4.44). Hence,

$$|\rho(t)|^2 \leqslant E_c\, E_H \qquad (4.46)$$

with the equality holding only if

$$H(f) = k\Psi^*(f)\, e^{-j2\pi ft} \qquad (4.47)$$

where k is an arbitrary scale factor. By the use of Eq. (4.44) we obtain

$$E_H = k^2\, E_c \qquad (4.48)$$

Since no loss in generality results if we set k = 1, let us make this assumption. In Eq. (4.47) we note that $H(f)$ is also a function of t, which can be treated as a parameter. Since we wish to obtain a single filter, let us arbitrarily set t = 0. Hence

$$H(f) = \Psi^*(f) \qquad (4.49)$$

where the impulse response is given by

$$h(t) = \psi^*(-t) \qquad (4.50)$$

which is the conjugate complex of the input waveform with the time axis reversed. This completes the derivation of the matched filter. The output response is obtained by the use of Eq. (4.29) as

$$\rho(t) = \int_{-\infty}^{\infty} \psi(\xi)\, \psi^*(\xi-t)\, d\xi \qquad (4.51)$$

which we denote as the *autocorrelation function* of $\psi(t)$. Usually the lag is denoted as τ and the autocorrelation function is written as

$$\rho(\tau) = \int_{-\infty}^{\infty} \psi(t)\, \psi^*(t-\tau)\, dt \qquad (4.52)$$

The autocorrelation function can be written in terms of the convolution. Using the shorthand notation in Eq. (4.31), we can write Eq. (4.52) as

$$\rho(\tau) = \psi(\tau) \star \psi^*(-\tau) \qquad (4.53)$$

In terms of the spectrum we can use Eqs. (4.40) and (4.49) to obtain

$$E(f) = |\Psi(f)|^2 \qquad (4.54)$$

which is designated as the *energy density spectrum* of $\psi(t)$, thus the choice of the symbol E. The autocorrelation function and energy density spectrum are Fourier transform pairs as

$$\rho(\tau) = \int_{-\infty}^{\infty} E(f)\, e^{j2\pi f\tau}\, df \qquad (4.55)$$

$$E(f) = \int_{-\infty}^{\infty} \rho(\tau)\, e^{-j2\pi f\tau}\, d\tau \qquad (4.56)$$

From Eqs. (4.54) and (4.55) the peak of the autocorrelation function is given by

$$\rho(0) = \int_{-\infty}^{\infty} E(f)\, df$$

$$= \int_{-\infty}^{\infty} |\Psi(f)|^2\, df$$

$$= E_c \qquad (4.57)$$

4.7/Random Noise

So far we have implicitly assumed that all signals are deterministic. We will now investigate the properties of random signals.

First of all, we will define an *ensemble average* as an average obtained by repeating the experiment with independent random data. It is not to be confused with a *time average*, which is performed over one run of the experiment. In general, the two averages can yield different results. If they yield the same result, the process is *ergodic*. For our purposes the ensemble average will play the dominant role in simulation. Next we postulate a random process $\{x(t)\}$ that is *stationary* such that the ensemble average

$$R(\tau) = \overline{x(t)x^*(t-\tau)} \qquad (4.58)$$

is a function of only the time difference τ and not the time origin.* We denote the ensemble average in Eq. (4.58) as the *autocorrelation function* of the random process $\{x(t)\}$. We will treat $x(t)$ as a complex function.

There are several useful properties of the autocorrelation functions of a stationary process. First of all, the autocorrelation function is symmetrical about the origin as

$$R(-\tau) = R^*(\tau) \qquad (4.59)$$

The ensemble averaged power in $\{x(t)\}$ is given by

$$P = \overline{|x(t)|^2} = R(0) \qquad (4.60)$$

---

*Such a condition usually is called *wide sense stationarity*.

If the random process is not periodic, the components in Eq. (4.58) will become uncorrelated as $\tau \to \infty$ and

$$\lim_{\tau \to \infty} R(\tau) = \overline{|x(t)|^2} \qquad (4.61)$$

Finally, by the use of the Schwartz inequality,

$$|R(\tau)| \leqslant R(0) \qquad (4.62)$$

Since $\{x(t)\}$ is stationary, it exists for all time and therefore has infinite energy. Consequently, the Fourier transform of $x(t)$ does not exist. However, if we restrict our interest to a finite length time interval, the energy will be finite in this interval and we can proceed to take the Fourier transform as

$$X_T(f) = \int_{-T/2}^{T/2} x(t)\, e^{-j2\pi ft}\, dt \qquad (4.63)$$

where we use the subscript T to emphasize the fact that the time interval is of finite length T.

In Eq. (4.54) we defined the energy density spectrum for a deterministic signal. We can do the same here for Eq. (4.63) in terms of the ensemble average as

$$E_T(f) = \overline{|X_T(f)|^2} \qquad (4.64)$$

However, this is a function of the time interval T as we can show by integration

$$E_T = \int_{-\infty}^{\infty} E_T(f)\,df = T \cdot P \qquad (4.65)$$

We introduced this finite time interval primarily just to take a Fourier transform. It would be preferable to define a spectral function that is independent of T. If we examine Eq. (4.65), we can accomplish this by dividing the energy spectral density by T to obtain a new function that has the dimensions of spectral power density. As it stands this new function would still be a function of T in the limits of the integration. To remove this dependence the *power spectral density* is defined by the limit

$$S(f) = \lim_{T\to\infty}\left[\frac{1}{T}\, E_T(f)\right] \qquad (4.66)$$

From Eq. (4.64) we note that the power spectral density is a nonnegative real function.

We can substitute Eqs. (4.63) and (4.64) into Eq. (4.66) to relate the power spectral density to the autocorrelation function. But since the derivation is somewhat tedious [Ref. 7], it will serve our purpose best just to give the result, which is

$$S(f) = \int_{-\infty}^{\infty} R(\tau)\, e^{-j2\pi f\tau}\, d\tau \qquad (4.67)$$

which is just a Fourier transform. We can complete the Fourier transform pair by writing

$$R(\tau) = \int_{-\infty}^{\infty} S(f)\, e^{j2\pi f\tau}\, df \qquad (4.68)$$

Note that the ensemble averaged power is given by

$$P = R(0) = \int_{-\infty}^{\infty} S(f)\, df \qquad (4.69)$$

Throughout the remainder of this book we will assume that all random processes will be stationary such that either the autocorrelation function or the power spectral density will completely specify them.

4.8/Filtering Random Noise

We can filter random noise in the same way as we can filter a deterministic signal. Let us designate the input random process as $\{x(t)\}$ where $R_x(\tau)$ and $S_x(f)$ are the autocorrelation function and power spectral density, respectively. The output of the filter can be written in the frequency domain as a product as

$$Y(f) = X(f)\, H(f) \qquad (4.70)$$

or in the time domain as a convolution as

$$y(t) = \int_{-\infty}^{\infty} h(t_1)\, x(t-t_1)\, dt_1 \qquad (4.71)$$

The autocorrelation function of $\{y(t)\}$ is

$$\begin{aligned}
R_y(\tau) &= \overline{y(t)\, y^*(t-\tau)} \\
&= \iint_{-\infty}^{\infty} h(t_1)h^*(t_2)\, \overline{x(t-t_1)x^*(t-\tau-t_2)}\, dt_1\, dt_2 \\
&= \iint_{-\infty}^{\infty} h(t_1)h^*(t_2)\, R_x(\tau+t_2-t_1)\, dt_1\, dt_2 \qquad (4.72)
\end{aligned}$$

After some manipulation, we can write this as a convolution of two functions as

$$R_y(\tau) = \int_{-\infty}^{\infty} R_h(t)\, R_x(\tau-t)\, dt \qquad (4.73)$$

where

$$R_h(\tau) = \int_{-\infty}^{\infty} h(t)\, h^*(t-\tau)\, dt \qquad (4.74)$$

We can also write a solution in terms of the power spectral densities $S_x(f)$ and $S_y(f)$. But first, we note that the Fourier transform of Eq. (4.74) is $|H(f)|^2$. Now the Fourier transform of Eq. (4.73) is the product of the spectral functions as

$$S_y(f) = S_x(f) \, |H(f)|^2 \qquad (4.75)$$

Note that the ensemble average properties of the random process $\{y(t)\}$ are not functions of the phase of $H(f)$ or the time origin of $h(t)$. While they have the same appearance, it is not straightforward to derive Eq. (4.75) directly from Eq. (4.70).

In many cases it is convenient to assume that a random process is broadband and spectrally flat. Such a process is designated as *white noise* where the power spectral density is constant over all frequencies as

$$S_x(f) = N_o \qquad (4.76)$$

By the use of Eq. (4.28) the autocorrelation function is

$$R_x(\tau) = N_o \delta(\tau) \qquad (4.77)$$

If we pass white noise through a filter, we obtain

$$S_y(f) = N_o \, |H(f)|^2 \qquad (4.78)$$

and

$$R_y(\tau) = N_o \int_{\infty}^{\infty} h(t) \, h^*(t-\tau) \, dt \qquad (4.79)$$

By the use of Eq. (4.69) we can write the ensemble averaged power in $\{y(t)\}$ as

$$P_y = N_o \int_{-\infty}^{\infty} |H(f)|^2 \, df = N_o \int_{-\infty}^{\infty} |h(t)|^2 \, dt$$
$$= N_o E_H \qquad (4.80)$$

where E_H is the energy in the complex function $h(t)$.

References

[1] Rihaczek, A.W.; *Principles of High-Resolution Radar*, Chapter 2; McGraw-Hill, New York, 1969.

[2] Papoulis, A.; *The Fournier Integral and Its Applications*; McGraw-Hill, New York, 1962.

[3] Bracewell, R.; *The Fourier Transform and Its Applications*; McGraw-Hill, New York, 1965.

[4] Champeney, D.C.; *Fourier Transforms and Their Physical Applications*; Academic Press, New York, 1973.

[5] Brigham, E.O.; *The Fast Fourier Transform*; Prentice-Hall, Englewood Cliffs, New Jersey, 1974.

[6] Papoulis, A.P.; *Probability, Random Variables, and Stochastic Processes*; McGraw-Hill, New York, 1965.

[7] Davenport, W.B. Jr., and W.L. Root; *An Introduction to the Theory of Random Signals and Noise*; McGraw-Hill, New York, 1958.

[8] McGillem, C.D., and G.R. Cooper; *Continuous and Discrete Signal and System Analysis*; Holt, Rinehart and Winston, New York, 1974.

Chapter 5

Received Signals and Receiver Responses

When we assume that scattering in the radar environment occurs at discrete scattering centers, it is straightforward to write the received radar signal as the superposition of the component signals given by Eq. (2.7). In this expression we have treated the motions of the scatterer and the antenna exactly. But while it is easy to write such an expression, it is not easy to compute the result. In this chapter we will derive expressions that can be more easily implemented, although their applicability has certain limitations, as was discussed in Section 2.3.

In this chapter we will also derive the receiver response to a collection of scatterers, where we assume that all processing in the receiver is linear. We sill see that the so-called ambiguity function plays an important role here. The five classes of waveforms will also be discussed in detail in this chapter.

5.1/Received Signals

The three time varying functions in Eq. (2.7) — $\tau(t)$, $G(t)$, and $\gamma(t)$ must be simplified in order to derive a computationally efficient expression for the received signal from a point scatterer. The latter two functions, the antenna modulation $G(t)$ and the fluctuating target $\gamma(t)$, are not usually rapidly varying functions. For most cases we can treat them as constants during the time interval of interest without losing any generality in the use of Eq. (2.7). For these cases, then, we can write the received signal as a delayed and rescaled version of the transmitted signal $\psi_T(t)$ as

$$\psi_R(t) = V_k \psi_T[t-\tau(t)] \qquad (5.1)$$

where $\tau(t)$ is the round trip delay and

$$V_k = \left[\frac{\lambda}{(4\pi)^{3/2}}\right] \frac{1}{r_k{}^2} G_k \gamma_k \qquad (5.2)$$

is a complex quantity associated with the kth point scatterer that takes into account range scaling $(1/r_k{}^2)$, the antenna pattern G_k, and the complex reflection coefficient γ_k of the scatterer. For convenience, we will designate V_k simply as the *range voltage coefficient* of the scatterer to distinguish it from the complex reflection coefficient. In terms of the complex modulation function $\mu_T(t)$ we can write

$$\psi_T(t) = \mu_T(t)e^{j2\pi f_c t} \qquad (5.3)$$

and

$$\psi_R(t) = V_k \mu_T[t-\tau(t)]e^{j2\pi f_c[t-\tau(t)]} \qquad (5.4)$$

The round trip delay $\tau(t)$ is a function of the time that depends on the target range at the instant of signal reflection (one-half the round trip delay) as

$$\tau(t) = \frac{2}{c} r[t-\tau(t)/2] \qquad (5.5)$$

where $r(t)$ is the target range as a function of time and c is the propagation velocity. To simplify the analysis we can perform Taylor series expansions on both $r(t)$ and $\tau(t)$. After some manipulation we can write

$$\tau(t) = \tau_k - \nu_k(t-t_o)/f_c + \dots \qquad (5.6)$$

where τ_k and ν_k are delay and Doppler coefficients of the kth scatterer given by Eqs. (2.9) and (2.10), respectively, and t_o is the instant of time at which τ_k and ν_k are defined. To derive Eq. (5.6) we have to assume that the relative velocity of the point scatterer is small compared to the propagation velocity (i.e., no relativistic effects); however, we will soon see that the relative velocity is subject to an even more restrictive constraint.

We are not yet ready to substitute Eq. (5.6) into Eq. (5.4). The result would still be too difficult to solve easily. There are two problems. The first is the stretching of the time scale in the complex modulation function as $t(1+\nu_k/f_c)$; in other words, the range to the point scatterer changes with time as $\tau_k - \nu_k t$. Since the range measurement is performed on the basis of the complex modulation function, the scatterer appears to be moving while the measurement is being performed. However, if the total change in range is small compared to the range resolution, there will be no loss in generality if we assume that the target is stationary for the purpose of the range measurement. Such a condition leads us to the bound on range rate in Eq. (2.12). The second problem is the presence of acceleration and higher order phase terms. If we can neglect these terms, the resulting received signal is easily processed. Such a condition leads us to the bound on range acceleration in Eq. (2.14). If the above two conditions are met, we can write the received signal from the kth point scatterer as

$$\psi_R(t) = V_k \mu_T(t-\tau_k)e^{j2\pi[f_c(t-\tau_k)+\nu_k(t-t_0)]} \quad (5.7)$$

Random Scatterers

If we use Eq. (5.7) to predict the received signal from a point scatterer, two difficulties arise. First, we must specify the complex quantity V_k when usually only scalar information is available for scatterers. Second, we must specify τ_k to a fraction of a wavelength in the phase term as can be demonstrated by writing

$$e^{-j2\pi f_c\tau_k} = e^{-j4\pi r_k/\lambda} \quad (5.8)$$

where r_k is the range coefficient of the kth scatterer. In practice this accuracy is not feasible. The best we can do is to treat the product, $V_k e^{-j4\pi r_k/\lambda}$, as a random phasor where the phase is uniformly distributed over 2π radians. With this assumption we will rewrite Eq. (5.7) as

$$\psi_R(t) = V_k \mu_T(t-\tau_k)e^{j2\pi(f_c+\nu_k)t} \quad (5.9)$$

where V_k now includes the range dependent phase term given by Eq. (5.8). We have also omitted the constant phase term, $\nu_k t_0$, since this will have no effect on the random signal. We see that three parameters now completely specify the point scatterer, namely V_k, τ_k, and ν_k (or equivalently, γ_k instead of V_k). Since V_k is random, we can write the ensemble average power in Eq. (5.9) as

$$\overline{|\psi_R(t)|^2} = \overline{|V_k|^2}\,|\mu_T(t-\tau_k)|^2 \quad (5.10)$$

The first factor on the right is designated as the *range power coefficient* of the scatterer, and from Eq. (5.2) it is given by

$$\overline{|V_k|^2} = \frac{\lambda^2}{(4\pi)^3}\frac{1}{r_k^4}G_k^2\overline{|\gamma_k|^2} \quad (5.11)$$

But from Eq. (2.3), the last quantity is just the ensemble average RCS associated with the kth scatterer. If we designate $\bar\sigma_k$ as this quantity, we obtain

$$\overline{|V_k|^2} = \frac{G_k^2\lambda^2}{(4\pi)^3 r_k^4}\bar\sigma_k \quad (5.12)$$

which relates the ensemble average range power coefficient of the scatterer to the conventional parameters in the radar range equation. System losses can also be included in the Eqs. (5.2) and (5.12). We can also include the transmitter power if we scale the transmitted signal $\mu_T(t)$ accordingly.

In choosing a value of $\bar\sigma_k$ to be used in Eq. (5.12) we must remember that the notation designates an ensemble average, which has been defined as an average obtained by repeating the experiment. It is not meant to be a time average; if it were, Eq. (5.10) would have no meaning.

Multiple Scatterers

By assuming that the scattering environment is composed of distinct point scatterers, each specified uniquely by a complex constant V_k and delay and Doppler coefficients τ_k and ν_k, the received signal is given by the superposition of the component responses in Eq. (5.9) as

$$\psi_R(t) = e^{j2\pi f_c t}\sum_k V_k\mu_T(t-\tau_k)e^{j2\pi\nu_k t} \quad (5.13)$$

Note that the carrier frequency comes outside of the summation. Since V_k is random, we can also derive the ensemble average power in Eq. (5.13), but a meaningful expression is obtained only if we assume that all scatterers are statistically independent. With this assumption we obtain

$$\overline{|\psi_R(t)|^2} = \sum_k \overline{|V_k|^2}\,|\mu_T(t-\tau_k)|^2 \quad (5.14)$$

which is independent of both f_c and ν_k.

While Eq. (5.13) is always valid for simulation purposes, it is inefficient to implement if the number of scatterers is large because the coefficients τ_k and ν_k (and also G_k in the definition of V_k) are likely to occur in a random or non-uniform manner. If the number of scatterers is large, it is preferable to regroup the scatterers on a rectangular grid before the computation in Eq. (5.13) is performed. We will show how this is done in Chapter 6.

5.2/ Linear Receiver Responses

It is common in radar receivers to design a filter that is matched to a signal $\psi_F(t)$, where this signal is either identical to the transmitted signal $\psi_T(t)$ or very close to it. In addition, Doppler filtering is usually employed in the receiver so that the general form of the signal to which the receiver is matched can be written as

$$\psi_F(t) = \mu_F(t)e^{j2\pi(f_c+\nu)t} \qquad (5.15)$$

where ν is the Doppler frequency to which the receiver is matched. The filter output for a single scatterer is given by the correlation of Eq. (5.15) with the received signal in Eq. (5.9) as

$$Z(\tau) = \int_{-\infty}^{\infty} \psi_R(t)\psi_F^*(t-\tau)dt \qquad (5.16)$$

After some manipulation we can write

$$Z(\tau) = V_k e^{j2\pi[(\nu_k-\nu)\tau_k+(f_c+\nu)\tau]} \int_{-\infty}^{\infty} \mu_T(t)\mu_F^*(t-\tau+\tau_k)e^{j2\pi(\nu_k-\nu)t}dt \qquad (5.17)$$

The response given by the integral portion of Eq. (5.17) can be written in terms of differential delay and Doppler as

$$\chi(\tau,\nu) = \int_{-\infty}^{\infty} \mu_T(t)\mu_F^*(t-\tau)e^{-j2\pi\nu t}dt \qquad (5.18)$$

This response is known as the matched-filter response or ambiguity function if $\mu_F(t)$ is identical to $\mu_T(t)$, or as the cross-ambiguity function if the two waveforms are not identical.* Point-target response and receiver response are terms that are also used. To avoid confusion, we will accept a somewhat liberal use of the term and define Eq. (5.18) as the *ambiguity function*, regardless of whether the two waveforms are identical or not.

The phase term outside of the integral in Eq. (5.17) is a constant and it does not affect the random properties of V_k. It is therefore expedient to combine it with V_k. We can now reduce Eq. (5.17) to

$$Z(\tau, \nu) = V_k \chi(\tau-\tau_k, \nu-\nu_k) \qquad (5.19)$$

Since the receiver response is also a function of ν, the Doppler mismatch, we have included it as an additional argument of Z. For multiple scatterers the response at the output of a linear receiver is given by the linear superposition of the component responses in Eq. (5.19) as

$$Z(\tau, \nu) = \sum_k V_k \chi(\tau-\tau_k, \nu-\nu_k) \qquad (5.20)$$

Since we have assumed V_k to be a random quantity, we can write the ensemble averaged power of Eq. (5.20) as

$$\overline{|Z(\tau,\nu)|^2} = \sum_k \overline{|V_k|^2}\,|\chi(\tau-\tau_k, \nu-\nu_k)|^2 \qquad (5.21)$$

where we have assumed that the scatterers are statistically independent.

If we examine the output of a linear receiver in Eq. (5.20) we observe that the order of events has been obscured somewhat by the linear superposition. The transmitted signal and the filtering process have been combined in the ambiguity

---

* Some definitions of the ambiguity function (e.g., Rihaczek [Ref. 2]) use a positive argument for the phase term in the integral in Eq. (5.18). It does not matter which convention is used as long as it is used consistently. The convention in Eq. (5.18) results in a more symmetrical form for Eq. (5.19).

function $\chi(\tau, \nu)$. The range voltage coefficient V_k acts as a simple scale factor. In fact, we can interpret Eq. (5.20) as the linear superposition of ambiguity functions, each placed at the delay-Doppler coordinate of a given scatterer and scaled by the range voltage coefficient of that scatterer. Notice that Eq. (5.20) is not an explicit function of the carrier frequency. The interpretation of Eq. (5.21) is identical to Eq. (5.20) except that power has replaced complex voltage. In Fig. 5.1 we illustrate how Eq. (5.21) works. In (a) we have a thumbtack ambiguity function $\chi(\tau, \nu)$; in (b) we have a collection of point scatterers in radar space representing $|V_k|^2$. The result of Eq. (5.21) is shown in (c). There is one more operation to implement in the receiver — detection. Various algorithms can be used. For example, if we implement square-law detection, the expression for the output of the detector is $|Z(\tau, \nu)|^2$. If the law is linear, the expression is $|Z(\tau, \nu)|$. Occasionally, operations are performed separately on the in-phase (I) and quadrature (Q) channels. In this case we can set $I = \mathrm{Re}\{Z(\tau, \nu)\}$ and $Q = \mathrm{Im}\{Z(\tau, \nu)\}$. For square-law detection the ensemble average of the detector output is given by Eq. (5.21).

Although we can use Eq. (5.20) to simulate the receiver response and Eq. (5.21) to predict the ensemble average response, the use of either expression is still inefficient to implement for a large number of scatterers. The same procedures that we will develop in Chapter 6 to regroup the scatterers onto a rectangular grid can also be used here to increase the efficiency of the simulation.

5.3/The Ambiguity Function

The ambiguity function has been defined by Woodward [Ref. 1] for the situation where the receiver is exactly matched to the transmitted waveform. But since receivers are seldom exactly matched in practice, it is convenient for our purposes to designate the ambiguity function as the response of the receiver to a point target, regardless of whether the receiver is matched or not. Thus the ambiguity function will be defined by Eq. (5.18). We can also write the ambiguity function in terms of the waveform spectra by substituting Eq. (4.16) into Eq. (5.18) for each waveform. After we rearrange the order of integration, we obtain

$$\chi(\tau, \nu) = \iint\limits_{-\infty}^{\infty} M_T(f_1)M_F{}^*(f_2)e^{j2\pi f_2\tau}\left\{\int\limits_{-\infty}^{\infty}e^{-j2\pi(\nu-f_1+f_2)t}dt\right\}df_1 df$$

(5.22)

The factor within the braces is just $\delta(\nu-f_1+f_2)$ according to Eq. (4.28) so that by setting $f_1 = f_2+\nu$ we can write

$$\chi(\tau, \nu) = \int\limits_{-\infty}^{\infty} M_T(f+\nu)M_F{}^*(f)e^{j2\pi f\tau}df$$

(5.23)

For the special case where the receiver is matched to the transmitted waveform as $\mu_F(t) = \mu_T(t) = \mu(t)$, the ambiguity function is given by

$$\chi(\tau, \nu) = \int\limits_{-\infty}^{\infty} \mu(t)\mu^*(t-\tau)e^{-j2\pi\nu t}dt$$

(5.24)

$$= \int\limits_{-\infty}^{\infty} M(f+\nu)M^*(f)e^{j2\pi f\tau}df$$

(5.25)

Figure 5.1/Result of Convolving Ambiguity Function with Collection of Point Scatterers in Radar Space.

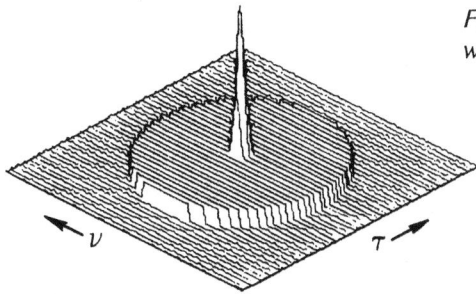

a. THUMBTACK AMBIGUITY FUNCTION

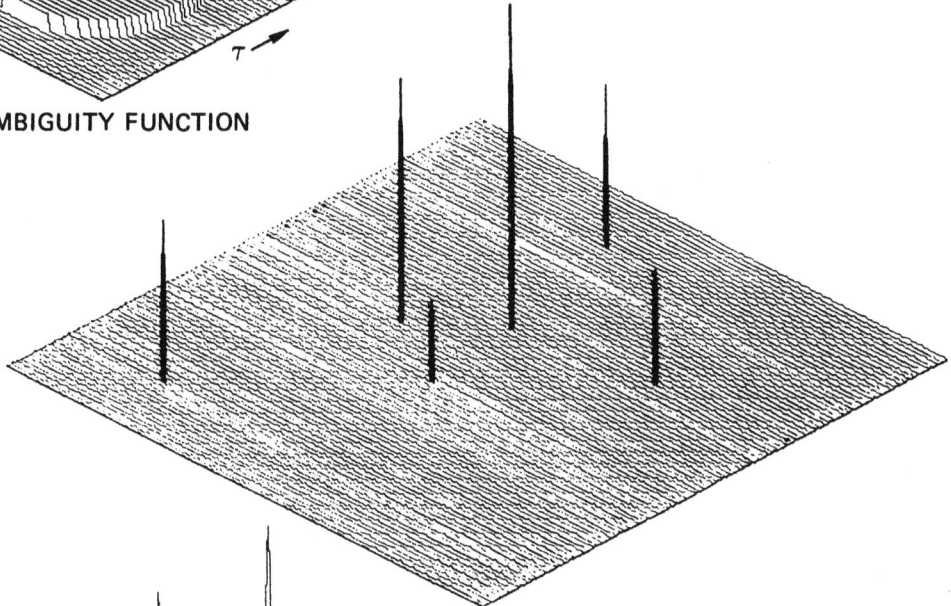

b. COLLECTION OF POINT SCATTERERS

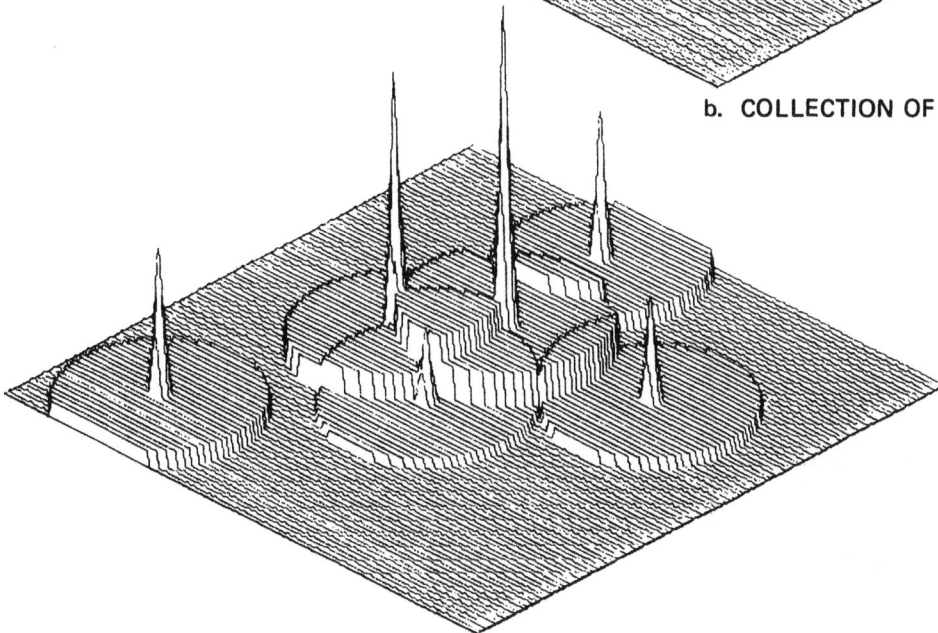

c. CONVOLUTION OF (a) WITH (b)

The peak response of a matched receiver occurs at the origin ($\tau = 0$, $\nu = 0$) and is given by

$$\chi(0, 0) = \int_{-\infty}^{\infty} |\mu(t)|^2 dt = \int_{-\infty}^{\infty} |M(f)|^2 df = E_c \qquad (5.26)$$

which is the energy in the complex signal. Now for the case of the mismatched receiver we can use the Schwartz inequality in Eq. (5.18) to obtain

$$|\chi(\tau, \nu)|^2 \leqslant \int_{-\infty}^{\infty} |\mu_T(t)|^2 \, dt \int_{-\infty}^{\infty} |\mu_F(t)|^2 dt$$

$$\leqslant E_T E_F \qquad (5.27)$$

where E_T and E_F are the energies in the complex signals $\mu_T(t)$ and $\mu_F(t)$, respectively. Receivers are usually mismatched only slightly and in a way that the peak of the response still occurs at the origin ($\tau = 0$, $\nu = 0$). Thus a convenient measure of the mismatch between the two waveforms is given by

$$\eta = \frac{|\chi(0, 0)|^2}{E_T E_F} = \frac{1}{E_T E_F} \left| \int_{-\infty}^{\infty} \mu_T(t)\mu_F{}^*(t)dt \right|^2 \qquad (5.28)$$

Note that if $\mu_T(t)$ is proportional to $\mu_F(t)$ then $\eta = 1$; otherwise $\eta < 1$.

An interesting property of the ambiguity function is that volume under the squared envelope is always constant as

$$\int_{-\infty}^{\infty}\int_{-\infty}^{\infty} |\chi(\tau, \nu)|^2 d\tau d\nu = \int_{-\infty}^{\infty} |\mu_T(t)|^2 dt \int_{-\infty}^{\infty} |\mu_F(t)|^2 dt$$

$$= E_T \cdot E_F \qquad (5.29)$$

Resolution

The ability of the receiver to resolve two point targets is determined by the ambiguity function in Eq. (5.18) or Eq. (5.23). Along the delay axis the ambiguity function is given by

$$\chi(\tau, 0) = \int_{-\infty}^{\infty} \mu_T(t)\mu_F{}^*(t-\tau)dt \qquad (5.30)$$

$$= \int_{-\infty}^{\infty} M_T(f)M_F{}^*(f)e^{j2\pi f\tau}df \qquad (5.31)$$

In Eq. (5.30) we recognize the expression as the cross-correlation function of the transmitted waveform with the waveform to which the receiver is matched. For modulated waveforms the resolution in delay is most conveniently determined from Eq. (5.31). Let us define B as the bandwidth of the narrowest spectrum in the integrand. If the band were rectangular, the Fourier transform in Eq. (5.31) would result in a $(\sin x)/x$ envelope in delay where the peak-to-null width of the response would be $1/B$. For the general case we simply define the resolution along the delay axis as

$$\Delta\tau = 1/B \qquad (5.32)$$

which is the definition used to obtain Eq. (2.11). If we want to define resolution more carefully (such as a half-power width or some other quantity), then we must perform the Fourier transformation in Eq. (5.31).

Along the Doppler axis the ambiguity function is given by

$$\chi(0, \nu) = \int_{-\infty}^{\infty} \mu_T(t)\mu_F{}^*(t)e^{-j2\pi\nu t}dt \qquad (5.33)$$

$$= \int_{-\infty}^{\infty} M_T(f+\nu)M_F{}^*(f)df \qquad (5.34)$$

For modulated waveforms it is now more convenient to work with Eq. (5.33) to determine resolution performance along the Doppler axis. Let us define T to be the shortest duration of the signals in the integrand of Eq. (5.33). Then from our previous arguments we can define the resolution along the Doppler axis as

$$\Delta\nu = 1/T \qquad (5.35)$$

which is the definition used in Eq. (2.13). We must remember that the quantities $1/B$ and $1/T$ are only estimates of the actual resolution capability of the receiver.

5.4/The Five Classes of Radar Waveforms

The waveform is the dominant feature of a radar system; depending on the waveform, we can resolve targets in range or motion, or both. Although there appear to be dozens of different waveforms discussed in the literature, they fall into only a handful of classes according to their resolution properties. Rihaczek [Ref. 3] defines three major classes — Class A, simple pulses; Class B, pulse compression waveforms; and Class C, coherent pulse trains. Among Class B waveforms, Rihaczek also distinguishes the linear-FM waveform as being unique. Although Rihaczek does not do so, it is expedient for simulation purposes to distinguish between long and short pulses in Class A waveforms. Therefore, we will consider only five classes of waveforms:

Class	Type
A1	Short Pulses
A2	Long Pulses (cw)
B1	Noise-Type Waveforms
B2	Linear-FM Waveforms
C	Coherent Pulse Trains

We will now discuss the properties of each class of waveforms. In addition we will show that the expressions earlier derived in this chapter can be simplified in all but one case (B1).

Class A1 — Short Pulses

Waveforms in this class will consist of a single pulse that is modulated only in amplitude. The bandwidth of such a pulse will be approximately the inverse pulse width as

$$B = 1/T_P \tag{5.36}$$

In other words, the time-bandwidth product of the pulse is about unity, $T_P B = 1$. Since resolution in delay is given by $\Delta\tau = 1/B$, we see that it is also determined by the pulse width as

$$\Delta\tau = T_P \tag{5.37}$$

In practical situations where T_P is short enough to achieve resolution in delay, it will not be possible to resolve scatterers in Doppler since the Doppler resolution is given by $\Delta\nu = 1/T_P$. Thus all scatterers can be assumed to be moving at the same Doppler velocity, which we will assume is zero for convenience. For example, the received signal and receiver response in Eqs. (5.13) and (5.20) can now be simplified as

$$\psi_R(t) = e^{j2\pi f_c t} \sum_k V_k \, \mu_T(t-\tau_k) \tag{5.38}$$

and

$$Z(\tau) = \sum_k V_k \, \chi(\tau-\tau_k) \tag{5.39}$$

where we note that $\chi(\tau)$ is the cross-correlation function of the two waveforms $\mu_T(t)$ and $\mu_F(t)$.

In most cases pulses are repeated in a pulse train. The processing in the receiver can be done in two ways — either coherently (processing the whole pulse train before envelope detection) or noncoherently (envelope detection on each pulse). It is far more efficient to simulate the pulse train as a whole rather than to implement Eq. (5.38) or Eq. (5.39) on each pulse. If the pulses are repeated at uniform intervals, then the methods developed for coherent pulse trains should be used even if the pulses are noncoherently processed. If the pulses are not repeated at uniform intervals (the so-called staggered pulse train), then the methods developed for noise-type waveforms should be used.

Class A2 — Long Pulses

Waveforms of this class will again consist of a single pulse that is modulated only in amplitude, but now the duration will be long enough for the scatterers to be resolved in Doppler. In a practical situation the pulses will be so long that it will not be possible to resolve scatterers in delay, since Eq. (5.37) applies also for long pulses. For simulation purposes we can assume that all scatterers will be at the same delay; thus, we can eliminate delay from the expressions derived earlier. Therefore the received signal and receiver response in Eqs. (5.13) and (5.20) can be simplified as

$$\psi_R(t) = \mu_T(t) \, e^{j2\pi f_c t} \sum_k V_k \, e^{j2\pi \nu_k t} \qquad (5.40)$$

and

$$Z(\nu) = \sum_k V_k \, \chi(\nu - \nu_k) \qquad (5.41)$$

where $\chi(\nu)$ is the spectrum of the product $\mu_T(t)\mu_F^*(t)$ as we can see from Eq. (5.18) with $\tau = 0$. It is also convenient to write Eq. (5.40) in terms of the signal spectrum as

$$\Psi_R(f) = \sum_k V_k M_T(f - f_c - \nu_k) \qquad (5.42)$$

where $M_T(f)$ is the spectrum corresponding to $\mu_T(t)$.

Class B1 — Noise-Type Waveforms

If the amplitude or phase modulation function of a pulse-compression waveform is irregular or noise-like, the ambiguity function will tend to consist of a sharp central spike on a wide pedestal of range-Doppler sidelobes, especially if the time-bandwidth product is large. Types of waveforms that are included in this category are:

Binary phase shift codes (pseudo-random codes, Barker codes, etc.)
Poly-phase codes (Frank codes, Huffman codes, etc.)
Nonlinear-FM pulses
Pulse trains with nonuniform or staggered PRF
Pulse trains with frequency shift coding
Long pulses with irregular amplitude modulation

For simulation purposes this class of waveforms is distinctive in that no simplifications are possible, in general, for the expressions derived earlier in this chapter. There are, however, two special cases.

The first is where the Doppler spread of all scatterers is low so that they cannot be resolved on the basis of Doppler — a common situation for applications where noise-type waveforms are used. For this case the simplifications obtained for the short pulse can also be used, where $\mu(t)$ is now the noise-type waveform.

The second case involves approximating the actual ambiguity function by the thumbtack, where the sidelobe pedestal represents the average sidelobe response. In addition, the sidelobe region is often much more extensive than the region containing scatterers, a fact that greatly simplifies the calculation of clutter power. The thumbtack approximation has application in analytical studies, in functional simulations, and in real-time simulations if the accuracy requirements are not so great.

Class B2 — Linear-FM Waveforms

Linear-FM waveforms are pulse-compression waveforms that are commonly used in practice because they are easy to implement; they are also easy to analyze and simulate. For all waveforms of this class the instantaneous frequency can be written as a linear function of time as

$$f_i(t) = Kt \qquad (5.43)$$

where K is designated as the slope of the FM, which can be either positive (up-slope) or negative (down-slope). The instantaneous phase of the signal is given by

$$\phi(t) = \pi K t^2 \qquad (5.44)$$

Let $a(t)$ designate the envelope of the signal. Then we can write the complex modulation function from Eq. (4.20) as

$$\mu(t) = a(t) \, e^{j\pi K t^2} \qquad (5.45)$$

If $a(t)$ is confined to a time duration T, the instantaneous frequency in Eq. (5.43) will sweep over a band of width $B = |K|T$. The instantaneous phase on a normalized time scale can also be specified uniquely in terms of the TB-product as

$$\phi(t) = \pm \pi T B (t/T)^2 \qquad (5.46)$$

where the choice in sign is given by the sign of the FM-slope.

If the envelope of the linear-FM signal is rectangular, the matched-filter response can be solved in closed form as

$$|\chi(\tau, \nu)| = (1 - |\tau|/T) \left| \text{sinc}[TB(1 - |\tau|/T)(\tau/T \mp \nu/B)] \right| \tag{5.47}$$

for $-T \leqslant \tau \leqslant T$ where $\text{sinc}(x) = (\sin \pi x)/\pi x$. The response is zero for $|\tau| > T$. In Fig. 5.2 we show the surface in Eq. (5.47) for TB = 25 and a down slope. The response is characterized by a ridge oriented along the line in the τ-ν plane given by $\tau/T = -\nu/B$. In practice, the high (-13 dB) side-lobes of the sinc function are undesirable, and some form of envelope tapering is used to reduce them [Ref. 2].

The linear-FM signal is unique among all waveforms in that its spectrum is also a linear-FM signal, at least for large TB-products. In fact the spectrum has approximately the same shape as the envelope in the time domain.

All linear-FM signals are characterized by a ridge in the ambiguity function, as in Fig. 5.2. Furthermore, there is strong symmetry in the ambiguity function about the ridge due to the linear coupling between time and frequency in the modulation function. A scatterer at a coordinate (τ_k, ν_k) will have about the same response as if it were located at $(\tau_k + \tau', \nu_k + \nu')$ for most values of τ' as long as

$$\nu' = K\tau' \tag{5.48}$$

The approximation is valid for values of τ' that are small compared to the signal duration T. We can use this result to simplify the expressions derived earlier for the received signal and receiver response. In Eq. (5.13) let us replace ν_k by $\nu_k + \nu'$ and τ_k by $\tau_k + \nu'/K$ to obtain

$$\psi_R(t) = e^{j2\pi f_c t} \sum_k V_k \mu_T(t - \tau_k - \nu'/K) e^{j2\pi(\nu_k + \nu')t} \tag{5.49}$$

Now if we set $\nu' = -\nu_k$, the phase term within the summation will drop out and we can write

$$\psi_R(t) = e^{j2\pi f_c t} \sum_k V_k \mu_T(t - \tau_k + \nu_k/K) \tag{5.50}$$

To obtain the received spectrum we should go back to Eq. (5.49) and set $\nu' = -K\tau_k$ and write

$$\psi_R(t) = e^{j2\pi f_c t} \sum_k V_k \mu_T(t) e^{j2\pi(\nu_k - K\tau_k)} \tag{5.51}$$

Then we can take the Fourier transform to obtain the received spectrum as

$$\Psi_R(f) = \sum_k V_k M_T(f - f_c - \nu_k + K\tau_k) \tag{5.52}$$

where $M_T(t)$ is the spectrum corresponding to $\mu_T(t)$.

To obtain the linear receiver response for the linear-FM signal let us replace ν_k by $\nu_k + \nu'$ and τ_k by $\tau_k + \nu'/K$ in Eq. (5.20) to obtain

$$Z(\tau, \nu) = \sum_k V_k \chi(\tau - \tau_k - \nu'/K, \nu - \nu_k - \nu') \tag{5.53}$$

Usually the receiver will be matched to zero Doppler ($\nu = 0$), so that if we set $\nu' = -\nu_k$ we can write

$$Z(\tau) = \sum_k V_k \chi(\tau - \tau_k + \nu_k/K) \tag{5.54}$$

where the reference to ν has now been omitted. The function $\chi(\tau)$ is just the cross-correlation function between $\mu_T(t)$ and $\mu_F(t)$. If the receiver processes in Doppler we can set $\nu' = -K\tau_k$ in Eq. (5.53) and reference delay to $\tau = 0$ as

$$Z(\nu) = \sum_k V_k \chi(\nu - \nu_k + K\tau_k) \tag{5.55}$$

The expressions in Eqs. (5.49) through (5.55) are valid as long as ν' is small compared to the bandwidth B, or equivalently

$$|\nu_k| \ll B \tag{5.56}$$

This condition will be valid in most simulation problems, but sometimes the Doppler frequencies will be too large for Eq. (5.66) to apply. However, in such cases the receiver will usually be matched to some mean Doppler frequency ($\nu = \nu_o$) so that $|\nu_k - \nu_o| \ll B$. In these cases we can set $\nu' = -(\nu_k - \nu_o)$ and proceed as before, or we can just define ν_k to be the residual Doppler frequency of the scatterer.

Class C — Coherent Pulse Trains

A waveform that has wide application for the detection of targets in clutter is the coherent pulse train — the only class of waveforms where it is possible to achieve a relatively clear area in both delay and Doppler about the central response peak of the ambiguity function.

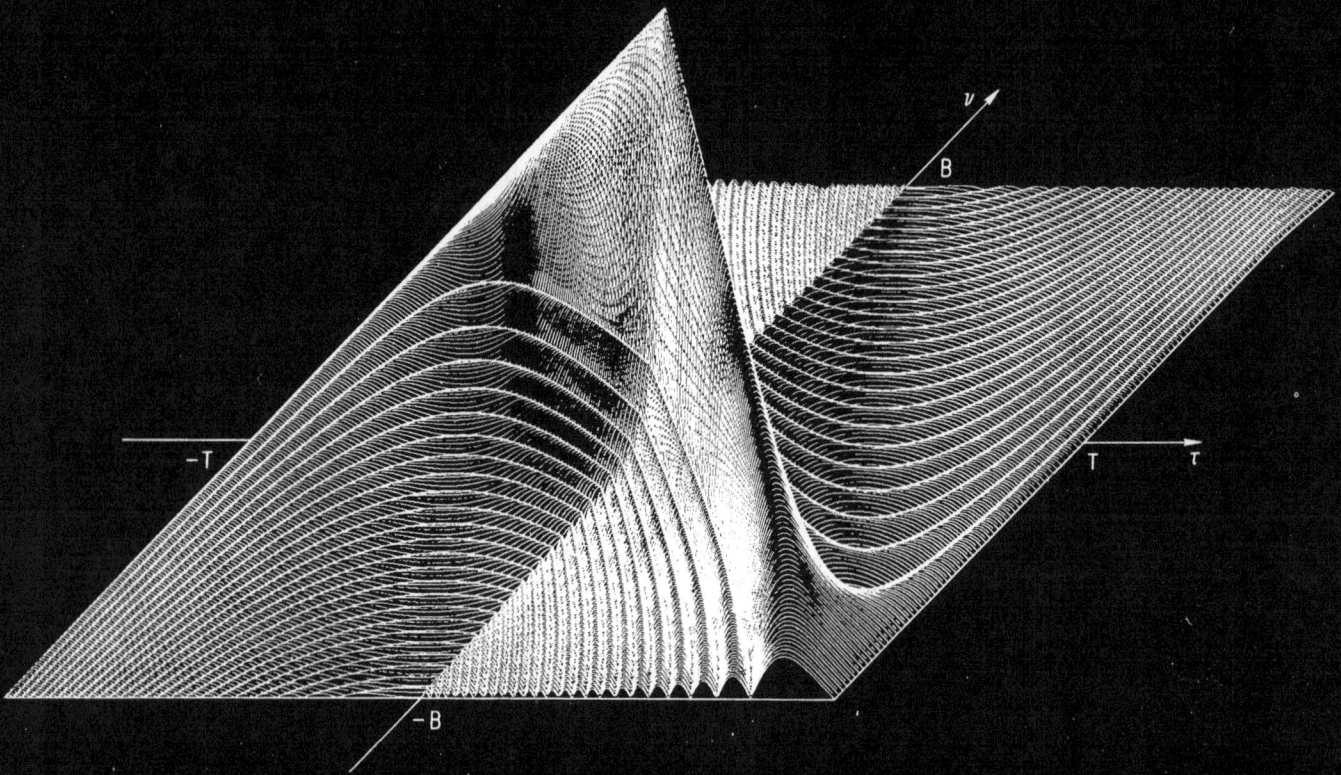

Figure 5.2/Envelope of Matched Filter Response of
Linear-FM Signal

The basic waveform consists of a sequence of
identical pulses repeated at uniform increments
in time. In general, each pulse can also be a
pulse compression waveform, as we show in
Fig. 5.3; therefore, we will write the complex
modulation function of the pulse train consisting of N pulses as

$$\mu_T(t) = \sum_{n=0}^{N-1} a_n \, \mu_{PT}(t-nT_r) \qquad (5.57)$$

where T_r is the *pulse repetition period*, and
$\mu_{PT}(t)$ is the complex modulation function of a
single pulse. We also have included a sequence
of complex weights $\{a_n\}$ in Eq. (5.57) to cover
three situations of interest. They are:

1) *Uniform coherent pulse train*
   Each pulse is uniformly weighted ($a_n = 1$)
   and the pulse train is completely coherent;
   therefore the values of $\mu_T(t)$ at the times
   $t = nT_r$ are all identical.*

---

* The concept of coherence in radar has nothing to do with the
definition of coherence in optics, which is a measure of how
narrow the radiation spectrum is. Complete coherence in optics implies a spectral line.

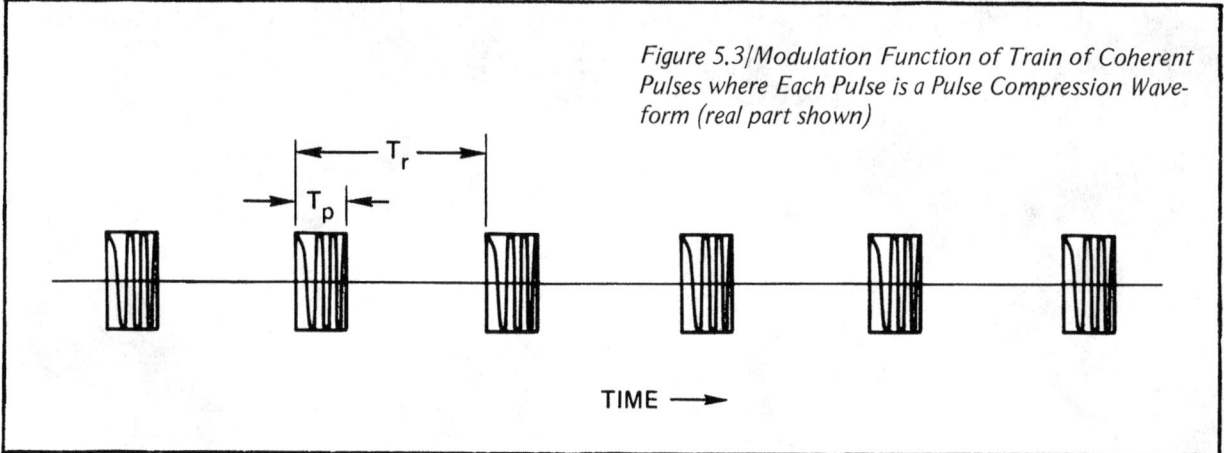

Figure 5.3/Modulation Function of Train of Coherent Pulses where Each Pulse is a Pulse Compression Waveform (real part shown)

2) *Weighted coherent pulse train*
   Each pulse can be weighted by a complex constant, but otherwise the train is completely coherent (predictable in a deterministic sense).

3) *Partially coherent or noncoherent pulse train*
   Each pulse is weighted inadvertently in the transmitter because of a lack of phase (and possibly amplitude) stability. Here $\{a_n\}$ can be treated as a random process.

In the following analysis we will assume that the receiver is matched to a waveform that is a slight variation of Eq. (5.57). The pulse waveform to which the receiver is matched is designated as $\mu_{PF}(t)$, and each pulse is weighted by a complex constant b_n as

$$\mu_F(t) = \sum_{n=0}^{N-1} b_n \mu_{PF}(t-nT_r) \qquad (5.58)$$

If we substitute Eqs. (5.57) and (5.58) into the definition of the ambiguity function in Eq. (5.18), we obtain

$$\chi(\tau, \nu) = \sum_{m=0}^{N-1}\sum_{n=0}^{N-1} a_n b_m{}^* \int_{-\infty}^{\infty} \mu_{PT}(t-nT_r)\mu_{PF}{}^*(t-mT_r-\tau)e^{-j2\pi\nu t}dt$$

$$= \sum_{m=0}^{N-1}\sum_{n=0}^{N-1} a_n b_m{}^* e^{-j2\pi\nu nT_r}\chi_P[\tau-(n-m)T_r, \nu] \qquad (5.59)$$

where $\chi_P(\tau, \nu)$ is the ambiguity function of a single pulse given by

$$\chi_P(\tau, \nu) = \int_{-\infty}^{\infty} \mu_{PT}(t)\mu_{PF}{}^*(t-\tau)e^{-j2\pi\nu t}dt \qquad (5.60)$$

Eq. (5.59) shows that the ambiguity function of a train of pulses consists of the superposition of the component ambiguity functions, translated to positions $\tau = (n-m)T_r$ in delay and weighted by a phase factor [Ref. 2 (Ch. 6)].

The appearance of the summations in Eqs. (5.57) through (5.59) is somewhat misleading, since at any given time in the pulse train there can be a contribution from only one component. To illustrate how this impacts on the ambiguity function, we have plotted in Fig. 5.4 the envelope of Eq. (5.59) for $N = 5$, $a_n = b_n = 1$, and where $\mu_{PT}(t)$ and $\mu_{PF}(t)$ are identical rectangular pulses of length $T_P = 0.3\ T_r$. We see that the response is confined to strips parallel to the Doppler axis. As long as the duty ratio does not exceed 50% (i.e., as long as $T_P \leqslant 0.5\ T_r$), the strips will not overlap. The strips are separated by T_r on the delay axis and within each strip we observe a repetition of peaks separated in Doppler by

$$f_r = 1/T_r \qquad (5.61)$$

which is designated as the *pulse repetition frequency*. For a 50% duty ratio or less we can write Eq. (5.59) as

$$\chi(\tau, \nu) = \chi_P(\tau-kT_r, \nu) \sum_n a_n b_{n-k}{}^* e^{-j2\pi\nu nT_r} \qquad (5.62)$$

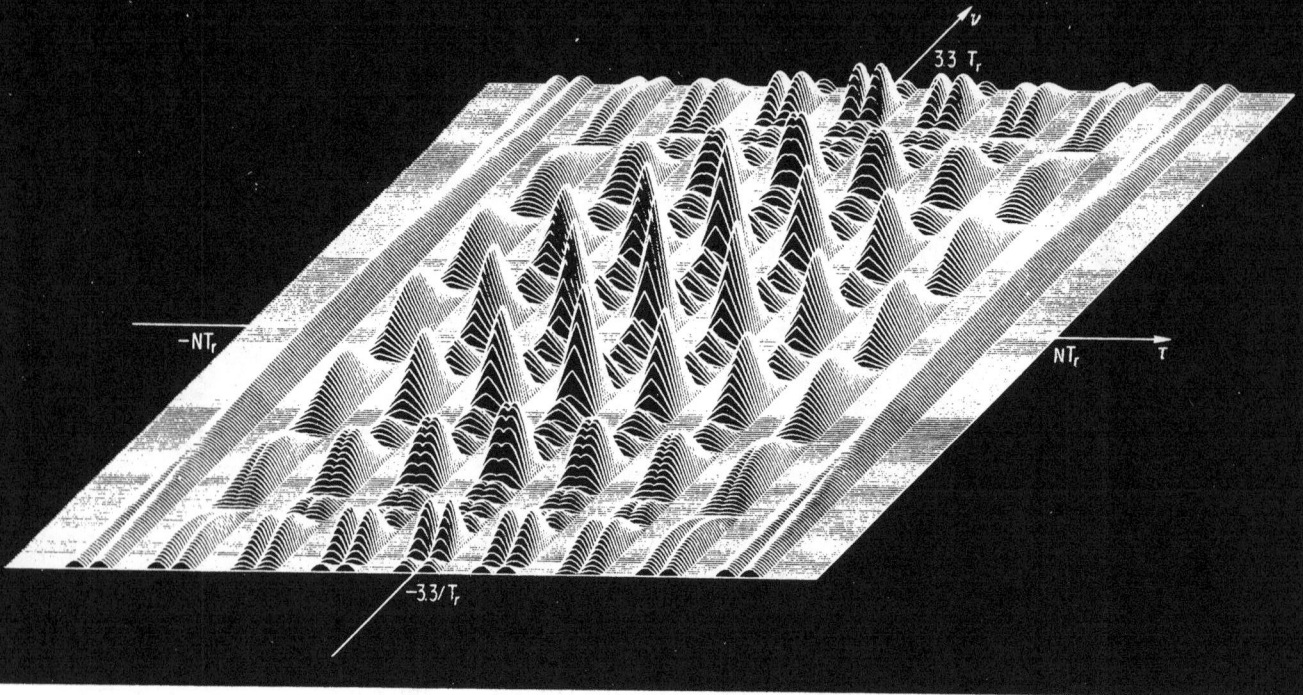

Figure 5.4/Envelope of Matched Filter Response for Coherent Pulse Train of 5 Rectangular Pulses

where
k = nearest integer to $[\tau/T_r]$

and the summation in n is performed over $k \leqslant n \leqslant N\text{-}1$ for $k \geqslant 0$ and $0 \leqslant n \leqslant N\text{-}1+k$ for $k < 0$. Of particular interest is the response near the origin (k = 0), which is given by

$$\chi(\tau, \nu) = \chi_P(\tau, \nu) \sum_{n=0}^{N-1} a_n b_n{}^* e^{-j2\pi\nu nT_r} \qquad (5.63)$$

There is only coarse Doppler resolution available on a single pulse. As long as $|\nu| \ll 1/T_P$, $\chi_P(\tau, \nu)$ deviates very little from $\chi_P(\tau, 0)$. Thus we can simplify Eq. (5.63) near the origin as the product of two functions as

$$\chi(\tau, \nu) = \chi_P(\tau)D(\nu) \qquad (5.64)$$

where $\chi_P(\tau)$ is the cross-correlation function between $\mu_{PT}(t)$ and $\mu_{PF}(t)$, and $D(\nu)$ is the *Doppler fine structure* given by

$$D(\nu) = \sum_{n=0}^{N-1} a_n b_n{}^* e^{-j2\pi\nu nT_r} \qquad (5.65)$$

If we set $\nu = \nu + m/T_r$ in Eq. (5.65) where m is any integer, we do not change the phase of the exponential. Thus $D(\nu)$ repeats with a repetition frequency given by Eq. (5.61).

In Fig. 5.4 we observe a matrix of peaks separated in delay by T_r and in Doppler by f_r. These peaks are known as the *ambiguities* of a coherent pulse train since it is not possible to distinguish a scatterer located at one peak from a different scatterer located at another peak. The peaks are nearly uniform in size, decreasing gradually from the origin in Doppler according to the Doppler resolution properties of a single pulse (the first factor in Eq. (5.62)), and in delay approximately as $1 - |k|/N$ where k is the order of the ambiguity (the number of terms being summed in Eq. (5.62)). In addition there is a slight difference in the nature of the Doppler sidelobes as we move from one strip to another. As long as the following two conditions are met

$$|\tau| \ll NT_r \qquad (5.66)$$

$$|\nu| \ll 1/T_P \tag{5.67}$$

we can approximate the ambiguous regions by the region surrounding the central peak as

$$\chi(\tau, \nu) = \chi_P(\tau - kT_r)D(\nu) \tag{5.68}$$

where

k = nearest integer to $[\tau/T_r]$

Sometimes more pulses are transmitted than are processed in the receiver. For this case Eq. (5.68) is exact, at least for small values of ν, as long as k does not exceed the number of extra pulses transmitted.

In most simulation problems we can use Eq. (5.68) to simplify the linear receiver responses in Eqs. (5.20) and (5.21). Since a scatterer at the coordinate (τ_k, ν_k) will have about the same response as if it were located at $(\tau_k \pm mT_r, \nu_k \pm nf_r)$, where m and n are any integers that are not too large, we can overlay all scatterers onto an area of size $T_r \times f_r$ (a region of unity time-bandwidth product). For convenience we will designate the working area as the area bounded by

$$0 \leqslant \tau < T_r \tag{5.69}$$

$$0 \leqslant \nu < f_r \tag{5.70}$$

Now Eqs. (5.20) and (5.21) can be written as

$$Z(\tau, \nu) = \sum_k V_k \chi_P(\tau - \tau_k')D(\nu - \nu_k') \tag{5.71}$$

and

$$|Z(\tau, \nu)|^2 = \sum_k |V_k|^2 |\chi_P(\tau - \tau_k')|^2 |D(\nu - \nu_k')|^2 \tag{5.72}$$

where

$$\tau_k' = \tau_k \;(\text{mod } T_r)$$

$$\nu_k' = \nu_k \;(\text{mod } f_r)$$

The mod or modulo operations are defined such that some integer multiples of T_r and f_r are added to, or subtracted from, τ_k and ν_k, respectively, so that $0 \leqslant \tau_k' < T_r$ and $0 \leqslant \nu_k' < f_r$. By using the same procedure we can also simplify the received signal in Eq. (5.13). Let us define t for the moment as the time measured from the transmission of the nth pulse and confine t as

$0 \leqslant t < T_r$. Then we can replace t in Eq. (5.13) by $t + nT_r$ and write

$$\psi_R(t+nT_r) = a_n e^{j2\pi f_c(t+nT_r)} \sum_k V_k \mu_{PT}(t-\tau_k')e^{j2\pi\nu_k'(t+nT_r)} \tag{5.73}$$

One more simplification can be made if we note that for the phase term involving ν, the differential phase $2\pi\nu_k' t$ over the duration of the pulse is bounded by $2\pi |\nu_k'| T_P \ll 2\pi$, as we will soon see. In general, we can drop the t in this phase term without any significant effect on the result. Thus Eq. (5.73) becomes

$$\psi_R(t+nT_r) = a_n e^{j2\pi f_c(t+nT_r)} \sum_k V_k \mu_{PT}(t-\tau_k')\, e^{j2\pi\nu_k'nT_r} \tag{5.74}$$

From Eq. (5.74) we can derive the signal at various stages in the receiver. For example, the received signal is usually mixed down to video (dc) so that we can write the received video signal as

$$\mu_R(t+nT_r) = a_n \sum_k V_k \,\mu_{PT}(t-\tau_k')e^{j2\pi\nu_k'nT_r} \tag{5.75}$$

If pulse compression is implemented next, then the output corresponding to the nth pulse is given by

$$Z_n(\tau) = \int_0^{T_r} \mu_R(t+nT_r)\mu_{PF}*(t-\tau)dt$$

$$= a_n \sum_k V_k\, \chi_P(\tau - \tau_k')e^{j2\pi\nu_k'nT_r} \tag{5.76}$$

Next, if Doppler filtering is implemented, we would weight each response in Eq. (5.76) by $\omega_n e^{-j2\pi\nu nT_r}$, where $\{\omega_n\}$ is the sequence of receiver weights, and sum over n as

$$Z(\tau, \nu) = \sum_{n=0}^{N-1} \omega_n Z_n(\tau)e^{-j2\pi\nu nT_r}$$

$$= \sum_k V_k\, \chi_P(\tau - \tau_k') \sum_{n=0}^{N-1} a_n\omega_n e^{-j2\pi(\nu-\nu_k')nT_r} \tag{5.77}$$

We recognize the summation over n as the Doppler fine structure in Eq. (5.65) if $b_n = \omega_n*$. Thus Eq. (5.77) reduces to Eq. (5.71).

If the processing in the receiver is noncoherent (no Doppler filtering), we will envelope detect after pulse compression. For the case of square-law detection we can write the absolute value squared of Eq. (5.76) as

$$|Z_n(\tau)|^2 = |a_n|^2 \left| \sum_k V_k \, \chi_P(\tau - \tau_k'') e^{j2\pi\nu_k'nT_r} \right|^2 \quad (5.78)$$

Usually the transmitter will be stable in amplitude so that $|a_n|^2 = 1$. Hence

$$|Z_n(\tau)|^2 = \left| \sum_k V_k \, \chi_P(\tau - \tau_k') e^{j2\pi\nu_k'nT_r} \right|^2 \quad (5.79)$$

Note that the presence of the phase term within the summation means that the detected outputs will fluctuate from pulse to pulse (corresponding to the index n) unless ν_k' is identical for all scatterers.

The simplifications derived in Eqs. (5.71) through (5.79) depend on several conditions being satisfied. The first is a duty ratio of 50% or less (although this specification is no longer a restriction if a single pulse is phase modulated with a time-bandwidth product larger than about 10 since there will not be much of a contribution in $\chi_P(\tau, \nu)$ for $|\tau| > 0.5 \, T_r$). The second condition is that strong clutter appearing in Doppler sidelobes should not be confined to a narrow band of Doppler frequencies (as ground clutter is to a ground based radar), because the sidelobe suppression varies from one ambiguous Doppler strip to another. Such clutter should be treated with more exact methods. Two more conditions are the inequalities imposed by Eqs. (5.66) and (5.67) which mean that the delay-Doppler spread of all scatterers should be confined as

$$|\tau_k| \ll NT_r \quad (5.80)$$

and

$$|\nu_k| \ll 1/T_P \quad (5.81)$$

Sometimes these two inequalities will not be satisfied; in such cases, the receiver will usually be matched to some coordinate (τ_o, ν_o) so that

$$|\tau_k - \tau_o| \ll NT_r \quad (5.82)$$

and

$$|\nu_k - \nu_o| \ll 1/T_P \quad (5.83)$$

If we confine τ_o and ν_o to be integral multiples of T_r and f_r, respectively, then we can still use Eqs. (5.71) and (5.72) by referencing τ to τ_o and ν to ν_o; moreover, we can use Eqs. (5.74) through (5.79) by starting the pulse count at τ_o. If none of the above conditions can be satisfied, then we can still treat the pulse train as if it were a noise-type waveform as far as the simulation is concerned.

5.5/Nonlinear Processing

The expressions developed in Section 5.2 are based on the assumption that all processing in the receiver is linear up to envelope detection. Many radar processors employ some nonlinear operation, such as analog-to-digital conversion or clipping, before correlation processing. In addition, there may be equipment instabilities and nonlinearities that are undesirable but nevertheless present. Often the nonlinear operations are designed so that the deviations from linearity are small. In these cases the linear analysis may be valid. However, some nonlinearities, such as hard-limiting, deviate so much from linearity that an analytical solution is not tractable.

Usually the only practical way to treat nonlinearities is to simulate them as they occur in the receiver processing sequence. Thus we begin with the received signal $\psi_R(t)$ in Eq. (5.13) or any of the simplified versions in Section 5.4. As written, $\psi_R(t)$ contains the carrier frequency. The next step in the simulation is to beat $\psi_R(t)$ down to the frequency where the nonlinearity takes place, which could be at IF or video. However, nonlinearities that occur at IF usually affect only the amplitude of the signal, not the phase. Thus we can beat $\psi_R(t)$ down to video in practically all cases for simulation purposes, even though the nonlinearity may occur at IF. We now write the received video signal as

$$\mu_R(t) = \psi_R(t)e^{-j2\pi f_c t} \qquad (5.84)$$

There are several types on nonlinear operations that are implemented in practice. We will now discuss the more common ones.

Two common types of nonlinear operations occur at IF. They are nonlinear amplification (e.g., logarithmic) and clipping. Both are used to limit the dynamic range of the signal and both can be implemented in the simulation as an operation on the magnitude of $\mu_R(t)$. For example, the clipped signal can be written as

$$\mu_c(t) = \begin{cases} \mu_R(t) & , \text{if } |\mu_R(t)| \leqslant c \\ \dfrac{c\,\mu_R(t)}{|\mu_R(t)|} , & \text{otherwise} \end{cases} \qquad (5.85)$$

where c is the clipping voltage level.

The most common nonlinear operation that occurs at video is analog-to-digital conversion. We classify it as a nonlinear operation if there is an insufficient number of bits to handle the dynam-

ic range of the signal. Nonlinear operations at video are usually performed on the quadrature components (the real and imaginary parts) of the video signal; moreover, the operation is performed separately on each component. In Fig. 5.5 we show the voltage transfer function of one method of implementing a 3-bit A/D converter. The quantization step is designated as q. It is common practice to clip the signal at IF prior to A/D conversion at a voltage level of q^{N-1}, where N is the number of bits, so that the dynamic range of the A/D converter is not exceeded. If clipping at IF were not implemented, phase distortions would result whenever the dynamic range of the A/D converter was exceeded.

A severe type of nonlinearity that is used in some inexpensive radar receivers is hardlimiting. Here the voltage in each of the quadrature components is quantized to one bit (hardlimiting is also known as one-bit processing) according to the sign of that component.

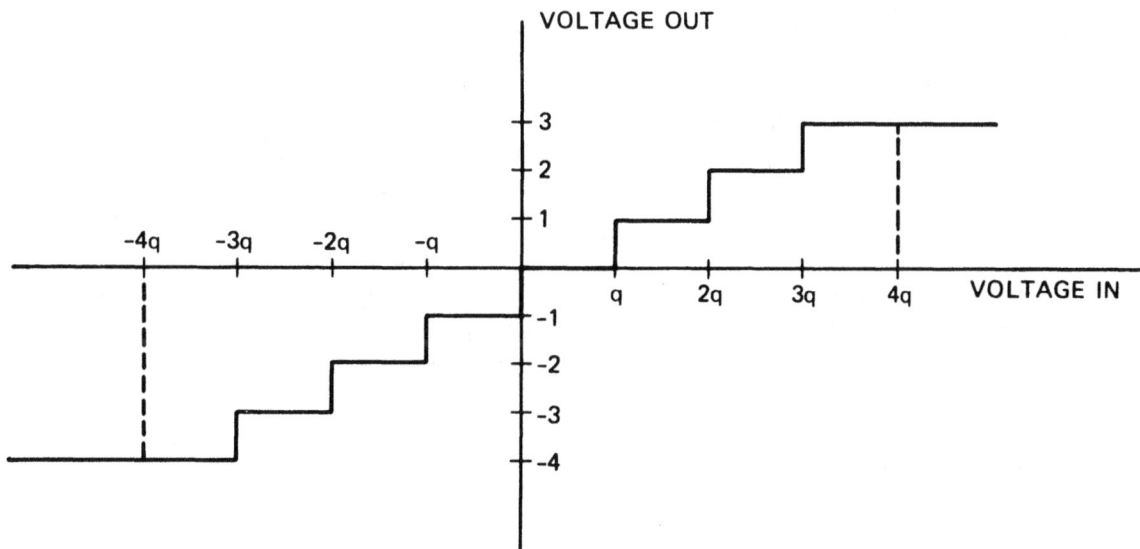

Figure 5.5/Voltage Transfer Function of 3-Bit A/D Converter

The nonlinear operation must be implemented on the received signal from all scatterers. In addition, the noise sources such as thermal noise and jamming must be included prior to the nonlinearity. We cannot utilize linear superposition on events that occur on opposite sides of the nonlinear operation.

5.6/Noise and Receiver Response to Noise

There are many sources of noise in a radar system. Antenna and receiver noise are the result of thermal effects and are broadband (white noise). Noise due to jammers can be broadband, spectrally shaped, or it can even have deterministic features (e.g., as a repeater jammer). Some atmospheric effects can also appear as noise to a radar. In practically all cases the noise will be additive to the received signal $\psi_R(t)$. Let us designate the noise process prior to processing in the receiver as $\{n(t)\}$ where $S_n(f)$ is the power spectral density. The composite signal prior to processing in the receiver is

$$\psi_{R+N}(t) = \psi_R(t) + n(t) \qquad (5.86)$$

In simulating a linear receiver we can process each component separately and superimpose the result. If the receiver is matched to the signal $\psi_F(t)$, the receiver response to noise is

$$N(\tau) = \int_{-\infty}^{\infty} n(t)\ \psi_F{}^*(t-\tau)\ dt \qquad (5.87)$$

From the derivations in Section 4.8 we can write the power spectral density of the random process $\{N(\tau)\}$ in terms of the one for $\{n(t)\}$ as

$$S_N(f) = S_n(f)\ |\Psi_F(f)|^2 \qquad (5.88)$$

where $\Psi_F(f)$ is the Fourier transform of $\psi_F(t)$. For the case of thermal or white noise where $S_n(f) = N_o$ for all f,

$$S_N(f) = N_o\ |\Psi(f)|^2 \qquad (5.89)$$

If we integrate over frequency, we obtain the ensemble average noise power after processing as

$$P_N = N_o E_F \qquad (5.90)$$

where E_F is the energy in the complex signal $\psi_F(t)$. With a coherent pulse train, the ensemble average noise power after pulse compression but prior to Doppler filtering is given by Eq. (5.90), except that the energy is that of a single pulse to which the receiver is matched, $\mu_{PF}(t)$, as

$$P_N = N_o E_{PF} \qquad (5.91)$$

After Doppler processing the ensemble average noise power is given by

$$P_N = N_o E_{PF} \sum_{n=0}^{N-1} |\omega_n|^2 \qquad (5.92)$$

where $\{\omega_n\}$ is the sequence of receiver pulse weights ($\omega_n = b_n{}^*$).

For receiver noise the power density N_o is given by

$$N_o = kT \cdot F \qquad (5.93)$$

where k is Boltzmann's constant, T is the receiver temperature ($kT = 4 \times 10^{-21}$ watt-sec for $T = 290°K$), and F is the receiver noise figure.

If system losses are not included in the various forms of the radar range equation as in Eq. (2.1) or Eq. (5.12), we can increase the noise power density by the system losses L as

$$N_o = kT\ F\ L \qquad (5.94)$$

so that the received signal power relative to noise is consistent.

References

[1]   Woodward, P.M.; *Probability and Information Theory, with Applications to Radar*; Pergamon Press, New York, 1953.

[2]   Rihaczek, A.W.; *Principles of High-Resolution Radar*; McGraw-Hill, New York, 1969.

[3]   Rihaczek, A.W.; "Radar Waveform Selection — A Simplified Approach," *IEEE Trans. Aerospace & Electronics Systems, Vol. AES-7*, pp. 1078-1086, November 1971.

Chapter 6

Mapping Procedures

So far, we have derived expressions to simulate received radar signals and receiver responses from a radar environment composed of a collection of point scatterers. But as we mentioned in Chapter 5, the expressions are not particularly well suited for ease in computation. They are satisfactory for a limited number of scatterers, but when the number of scatterers is large, as with distributed clutter, they are not at all suitable. In this chapter we will derive mapping procedures that are especially efficient for extended targets and distributed clutter. The methods will not only be computationally fast, but they will also be easy to implement.

6.1 / Rectangular Gridded Formats

Although Eqs. (5.13), (5.20), and (5.21) give the desired responses to a collection of point scatterers, they are inefficient to use in numerical computations since the coordinates (τ_k, ν_k) are likely to occur in an irregular and unpredictable manner. We would either have to compute the complex modulation function $\mu_r(t)$ in Eq. (5.13) at each point of interest (a tedious computation) or employ some interpolation scheme if a matrix of values were already available (requiring excessive storage if more than one dimension is involved). It would be preferable to have a structure based on uniformly spaced samples, even though the scatterers themselves may never be arranged in such a structure. In two dimensions, this structure is a rectangular grid.

Let us now examine Eqs. (5.13) and (5.20). Suppose we have two nearby scatterers that are unresolvable. We could add the complex scattering

coefficients of each scatterer to create a new scatterer without significantly affecting the computation in either equation. To carry this procedure one step further, let us construct a rectangular cell of size $\Delta\tau$ by $\Delta\nu$ and add all scatterers within this cell to form a new composite scatterer. The only restriction will be that any two scatterers within this cell will be unresolvable. The composite range voltage coefficient for the mnth cell on a rectangular grid is now given by

$$V(\tau_m, \nu_n) = \sum_{k \in \Delta_{mn}} V_k \qquad (6.1)$$

where the summation is performed over all scatterers within the mnth cell. We sketch how this mapping is done in Fig. 6.1.

With Eq. (6.1) we can rederive the received signal corresponding to Eq. (5.13) as

$$\psi_R(t) = e^{j2\pi f_c t}\sum_m \sum_n V(\tau_m, \nu_n)\, \mu_T(t-\tau_m)e^{j2\pi\nu_n t} \qquad (6.2)$$

and the receiver response corresponding to Eq. (5.20) as

$$Z(\tau, \nu) = \sum_m \sum_n V(\tau_m, \nu_n)\, \chi(\tau-\tau_m, \nu-\nu_n) \qquad (6.3)$$

If we assume that all scatterers are independent, we can write the ensemble averaged power of Eq. (6.1) as

$$\overline{|V(\tau_m, \nu_n)|^2} = \sum_{k \in \Delta_{mn}} \overline{|V_k|^2} \qquad (6.4)$$

which is just a scalar addition. Thus we can write the ensemble averaged power in the receiver response corresponding to Eq. (5.21) as

$$\overline{|Z(\tau, \nu)|^2} = \sum_m \sum_n \overline{|V(\tau_m, \nu_n)|^2}\, |\chi(\tau-\tau_m, \nu-\nu_n)|^2 \qquad (6.5)$$

The expressions in Eqs. (6.2), (6.3), and (6.5) appear to be more complicated than their original counterparts because double summations are involved instead of single ones. However, we

*Figure 6.1/Procedure for Obtaining Rectangular Gridded
Format*

**WE BEGIN WITH A COLLECTION OF POINT-SCATTERERS AT ARBITRARY
POSITIONS IN THE τ - ν PLANE**

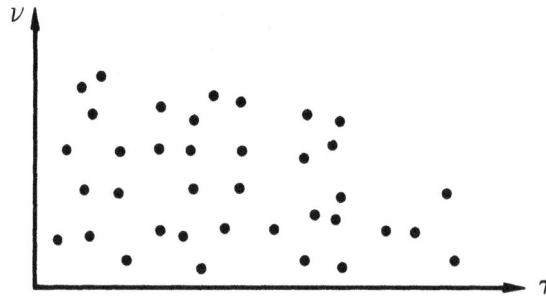

WE SUPERIMPOSE A RECTANGULAR GRID ON TOP OF THIS PLANE AS

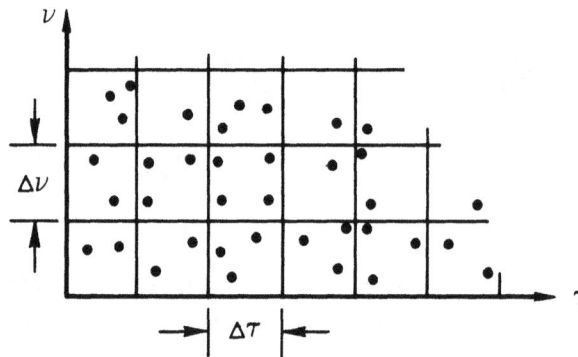

**WITHIN EACH CELL, WE COMBINE ALL SCATTERERS TO CREATE A NEW
SCATTERER AT THE CENTER OF THE CELL AS**

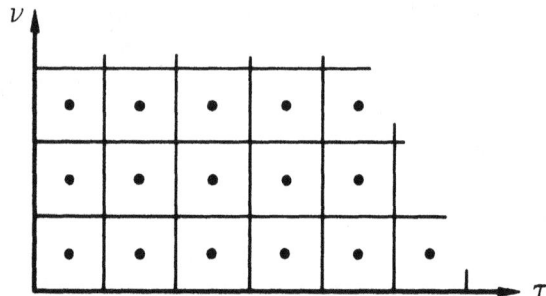

can recognize Eqs. (6.3) and (6.5) as two-dimensional discrete convolutions, and Eq. (6.2) as a combined discrete convolution and discrete Fourier transform. If we denote Eq. (6.4) as the distribution of the range power coefficient, then the receiver response is the convolution of this coefficient with the ambiguity function of the waveform. These discrete computations are greatly more efficient to implement on a rectangular grid (possibly several orders of magnitude). In addition, we note that by the use of either Eq. (6.1) or Eq. (6.4), the number of scatterers has actually been reduced (but in some cases, there may be more cells on the rectangular grid than actual scatterers; thus, many of the cells will be vacant). Finally, there are fast algorithms available to perform the discrete convolutions indirectly, as we will discuss in Section 7.6.

In Section 5.4 we discussed the simplifications that are possible for four of the five waveform classes. In Fig. 6.2 we show how Eqs. (6.2) and (6.3) can be written for the specific waveform classes. For short pulses (Class A1), the gridded format is just a sequence of strips parallel to the Doppler axis, where each strip is of uniform width $\Delta \tau$. The quantity $V(\tau_m)$ is the phasor sum of all range voltage coefficients falling within the mth cell centered at the delay coordinate τ_m. For long pulses (Class A2), we have the reverse situation. The strips of width $\Delta \nu$ are now parallel to the delay axis and $V(\nu_n)$ is the phasor sum of all range voltage coefficients falling within the nth cell centered at the Doppler coordinate ν_n. For noise type waveforms (Class B1), there are no further simplifications possible beyond Eqs. (6.2) and (6.3).

The simplified expressions for the linear-FM signal (Class B2) have the same appearance as those for short pulses, except the cell is now sheared so that it is parallel to the line $\tau = \nu / K$, where K is the slope of the linear FM. The cell width along the delay axis is $\Delta \tau$; $V(\tau_m)$ is the phasor sum of all range voltage coefficients falling within the mth cell, the center of which intersects

the delay axis at τ_m. Because of the linear coupling between time and frequency for the linear-FM signal, we could alternately use the simplified frequency-domain expressions for the long pulses if we designate ν_n as the Doppler coordinate where the center of the sheared cell intersects the Doppler axis.

The procedure for coherent pulse trains (Class C) appears complicated, but is actually straightforward. First of all, we must establish the rectangular grid within the unity TB-product area. Since we have some choice in the size of $\Delta \tau$ and $\Delta \nu$, we will choose them so that there will be an integral number of increments within T_r and f_r, respectively, as $N_\tau = T_r / \Delta \tau$ and $N_\nu = f_r / \Delta \nu$. The next step involves mapping all ambiguous cells onto the unity TB-product area. Thus $V(\tau_m, \nu_n)$ is the phasor sum of not only all range voltage coefficients that fall within the mn^{th} cell centered at $(\tau_m + T_r, \nu_n)$, $(\tau_m, \nu_n + f_r)$, etc. This mapping step has the added benefit of reducing the size of subsequent discrete convolutions and discrete Fourier transforms. Eqs. (5.71) and (5.74) can now be written in terms of the definite summations as shown in Fig. 6.2 (we have replaced the index n in Eq. (5.74) by ℓ in the figure). Because of the modulo operations involved in Eqs. (5.71) and (5.74), the discrete convolutions and discrete Fourier transforms are *circular*, a property we will define and discuss in the next chapter.

All we have done in this section and in Fig. 6.2 is to rewrite the expressions derived in Section 5.4 in a rectangular gridded format in order to be able to implement more efficient computational algorithms. Nothing has changed regarding the validity of the original expressions. Therefore, the five cases described in Fig. 6.2 are still subject to the conditions specified in Section 5.4.

Let us now go back to Eq. (6.4). We recognize this quantity as the RCS of the mn^{th} cell modified as in Eq. (5.12) by the radar range equation.

Thus we can write

$$\overline{|V(\tau_m, \nu_n)|^2} = \frac{G^2 \lambda^2}{(4\pi)^3 r^4} \sigma_{mn} \tag{6.6}$$

where G and r are the nominal antenna gain and range to the cell (assuming that all scatterers in the cell are weighted by the same antenna gain). Usually we will be given only RCS, so the scalar addition in Eq. (6.4) would be much simpler for simulation purposes than generating individual random phasors and performing a phasor addition in Eq. (6.1). We can still create a new random phasor $V(\tau_m, \nu_n)$ given the ensemble average power, but the phasor will, in general, have a different probability distribution function associated with it than the distribution function associated with the component variates in Eq. (6.1). This difference is a potential problem, as we discussed in Section 3.3, but, as we also pointed out, there are two limiting cases of interest that are tractable. The first case is the many-scatterer mechanism which leads to Rayleigh-amplitude statistics; the other is where one scatterer in a resolution cell dominates all other scatterers and the statistics of the composite are determined primarily by the dominant component. For either case, we can generate a random phasor $V(\tau_m, \nu_n)$ where the average amplitude is the squareroot of Eq. (6.6) and the phase is random (in Chapter 9 we show how this process can be implemented). If neither the many-scatterer mechanism nor the dominant-scatterer theory is valid, then the random phasors must be generated separately and summed in Eq. (6.1).

6.2/The Extra Dimension(s)-Angle

Let us substitute Eq. (5.2) into Eq. (6.1) to obtain

$$V(\tau_m, \nu_n) = \alpha_m \sum_{k \in \Delta_{mn}} G_k \gamma_k \tag{6.7}$$

where α_m is a range-dependent scale factor given by

$$\alpha_m^2 = \frac{\lambda^2}{(4\pi)^3 r_m^4} \tag{6.8}$$

Note that the antenna pattern is applied to each scatterer within the summation of Eq. (6.7). If the antenna is scanning as a part of the simulation, different weights will be applied to each scatterer within the summation for each different position of the antenna. The expressions derived in Section 6.1 are not computationally efficient for scanning antennas; however, if we extend the rectangular gridded structure to angle, we can derive expressions that are.

In Section 6.1, we grouped scatterers that were basically unresolvable. To extend this to a third dimension, we will collect the complex reflection coefficients that fall within the cell in 3-D space with the indices (ℓ, m, n) and define

$$\gamma(\theta_\ell, \tau_m, \nu_n) = \sum_{k \in \Delta_{\ell mn}} \gamma_k \tag{6.9}$$

All scatterers within the angle cell at θ_ℓ will be weighted uniformly by $G(\theta_\ell)$, the one-way antenna power gain in the direction defined by θ_ℓ. Just as in Section 6.1, θ_ℓ will be defined over a grid of uniformly spaced increments. Since all

Figure 6.2/Gridded Formats for Five Waveform Classes

CLASS A1: SHORT PULSES

$$\psi_R(t) = e^{j2\pi f_c t} \sum_m V(\tau_m)\mu_T(t-\tau_m)$$

$$Z(\tau) = \sum_m V(\tau_m)\chi(\tau-\tau_m)$$

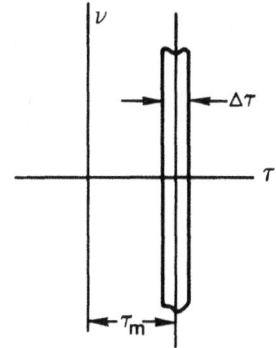

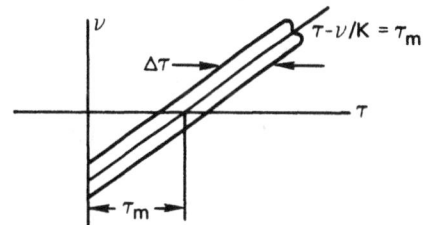

CLASS A2: LONG PULSES

$$\Psi_R(f) = \sum_n V(\nu_n)M_T(f-f_c-\nu_n)$$

$$Z(\nu) = \sum_n V(\nu_n)\chi(\nu-\nu_n)$$

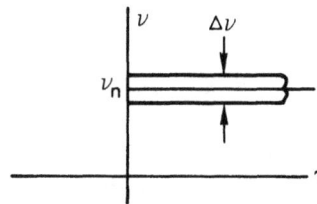

CLASS B1: NOISE-TYPE WAVEFORMS (GENERAL CASE)

$$\psi_R(t) = e^{j2\pi f_c t} \sum_m \sum_n V(\tau_m, \nu_n)\mu_T(t-\tau_m)e^{j2\pi\nu_n t}$$

$$Z(\tau, \nu) = \sum_m \sum_n V(\tau_m, \nu_n)\chi(\tau-\tau_m, \nu-\nu_n)$$

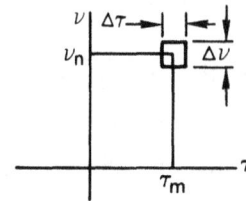

CLASS B2: LINEAR-FM

$$\psi_R(t) = e^{j2\pi f_c t} \sum_m V(\tau_m)\mu_T(t-\tau_m)$$

$$Z(\tau) = \sum_m V(\tau_m)\chi(\tau-\tau_m)$$

CLASS C: COHERENT TRAINS

$$\psi_R(t+\ell T_r) = a_\ell e^{j2\pi f_c(t+\ell T_r)} \sum_{m=0}^{N_\tau-1} \sum_{n=0}^{N_\nu-1} V(\tau_m, \nu_n)\mu_{PT}(t-\tau_m)e^{j2\pi\nu_n \ell T_r}$$

$$Z(\tau, \nu) = \sum_{m=0}^{N_\tau-1} \sum_{n=0}^{N_\nu-1} V(\tau_m, \nu_n)\chi_P(\tau-\tau_m)D(\nu-\nu_n)$$

$$N_\tau = T_r/\Delta\tau$$

$$N_\nu = f_r/\Delta\nu$$

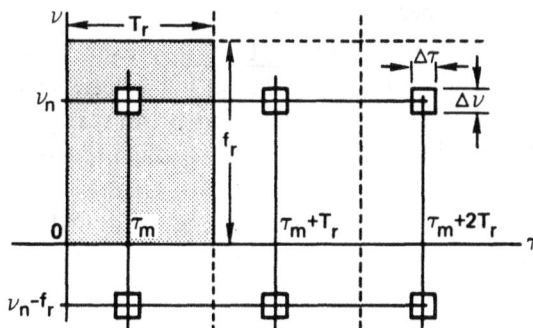

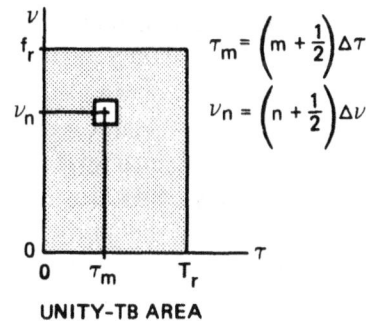

$$\tau_m = \left(m+\tfrac{1}{2}\right)\Delta\tau$$

$$\nu_n = \left(n+\tfrac{1}{2}\right)\Delta\nu$$

UNITY-TB AREA

signals and receiver responses are now a function of where the antenna is pointing, let us designate this angle as θ and the gain in the θ_ℓ direction as $G(\theta_\ell - \theta)$. Thus we can extend Eqs. (6.2) and (6.3) to the third dimension as

$$\psi_R(\theta, t) =$$
$$e^{j2\pi f_c t} \sum_\ell \sum_m \sum_n \alpha_m \gamma(\theta_\ell, \tau_m, \nu_n)\, G(\theta_\ell - \theta)\, \mu(t - \tau_m)\, e^{j2\pi\nu_n t} \tag{6.10}$$

and

$$Z(\theta, \tau, \nu) =$$
$$\sum_\ell \sum_m \sum_n \alpha_m \gamma(\theta_\ell, \tau_m, \nu_n)\, G(\theta_\ell - \theta)\, \chi(\tau - \tau_m, \nu - \nu_n) \tag{6.11}$$

If we assume that all scatterers are independent, then we can write the ensemble average power associated with Eq. (6.9) as

$$\overline{|\gamma(\theta_\ell, \tau_m, \nu_n)|^2} = \sum_{k \in \Delta_{\ell mn}} \overline{|\gamma_k|^2} \tag{6.12}$$

but this is just an expression for average RCS. Therefore, let us define

$$\overline{\sigma(\theta_\ell, \tau_m, \nu_n)} = \sum_{k \in \Delta_{\ell mn}} \overline{\sigma_k} \tag{6.13}$$

as being equivalent to Eq. (6.12). Now we can write the ensemble average receiver response in Eq. (6.11) as

$$\overline{|Z(\theta, \tau, \nu)|^2} =$$
$$\sum_\ell \sum_m \sum_n \alpha_m{}^2\, \overline{\sigma(\theta_\ell, \tau_m, \nu_n)} \cdot G^2(\theta_\ell - \theta)\, |\chi(\tau - \tau_m, \nu - \nu_n)|^2 \tag{6.14}$$

The expressions derived so far in this section only apply strictly to a fan-beam antenna where θ is measured in a plane normal to the fan-beam. In general, the antenna pattern will be two-dimensional. Thus we can extend these results to a two-dimensional scan by defining θ as a 2-D vector. The total radar space is then four-dimensional and the summations are four-fold. Even if the scan is only in one dimension, we cannot eliminate the four-fold summation for a 2-D antenna pattern unless the pattern is separable as

the product of one-dimensional functions. Then in the direction normal to the scan, the function for that coordinate is applied at the time the scatterers are collected in the cells in Eq. (6.9) or Eq. (6.13). The remaining function is then used in the three-fold summation.

The principal reason why the expressions derived in this section are efficient is that all three (or four) summations involve standard computational operations — namely discrete convolutions, discrete correlations, and discrete Fourier transforms. These operations are efficient only if the increments are uniform. For those cases where it is not feasible to have uniform increments in angle, then the expressions in Section 6.1 should be used for each angle of interest.

Although the expressions derived here are computationally efficient, three- and four-fold summations are still going to require a vast amount of computing resources if the radar space is extensive. Fortunately, we will not often be required to implement three-fold summations, let alone the four-fold variety. On the other hand, we will seldom get by with implementing anything less than a two-fold summation.

6.3/Mapping Onto the Grid

If the scattering models are expressed in radar space, then the ensemble average RCS $\sigma(\tau_m, \nu_n)$ or $\sigma(\theta_\ell, \tau_m, \nu_n)$ will be specified directly by the models. But if the scattering models are expressed in physical space, then scatterers must be mapped to radar space. Two operations are involved here — a coordinate transformation and the collection of scatterers into cells as in Eqs. (6.4) or (6.13). The techniques used to perform both operations are of extreme importance in simulation, because both computation time and accuracy are affected. In this section, we will discuss two methods of mapping scatterers onto the grid.

The first method is that implied by Eqs. (6.4) and (6.13) — the average power associated with each scatterer is accumulated at the center of the cell in which it is located, as indicated by Fig. 6.1. We will designate this approach as the *nearest-sample approach* to distinguish it from a slightly more complicated method. To increase the accuracy of the simulation, we can share the power associated with a scatterer among neighboring elements on the grid. This method we designate as the *shared-power approach*. A scatterer falling on a matrix element will get mapped into that element, but otherwise most of the power will go to the nearest element with proportionally less going to the remaining elements, as we show in Fig. 6.3. In one dimension, there will be two matrix elements receiving power; in two dimensions, there will be four; in three dimensions, eight; etc.

Power is shared among the matrix elements in a linear manner which is best described as a type of reverse interpolation. In one dimension, a scatterer located at a radar coordinate u (which could be any of the radar coordinates) will get mapped into two matrix (array) elements. The index of the first element is

m = largest integer less than $[(u-u_o)/\Delta u]$ (6.15)

where u_o is a coordinate reference and Δu is the element spacing. If the ensemble average RCS of the scatterer is designated as $\bar{\sigma}$, the amount being mapped into each element is

$$\left.\begin{array}{l} \bar{\sigma}_m = (1-\alpha)\bar{\sigma} \\ \sigma_{m+1} = \alpha\bar{\sigma} \end{array}\right\}$$ (6.16)

where α is the fractional distance given by

$\alpha = (u-u_o)/\Delta u - m$ (6.17)

In two or more dimensions, the expressions become proportionately more complex. For ex-

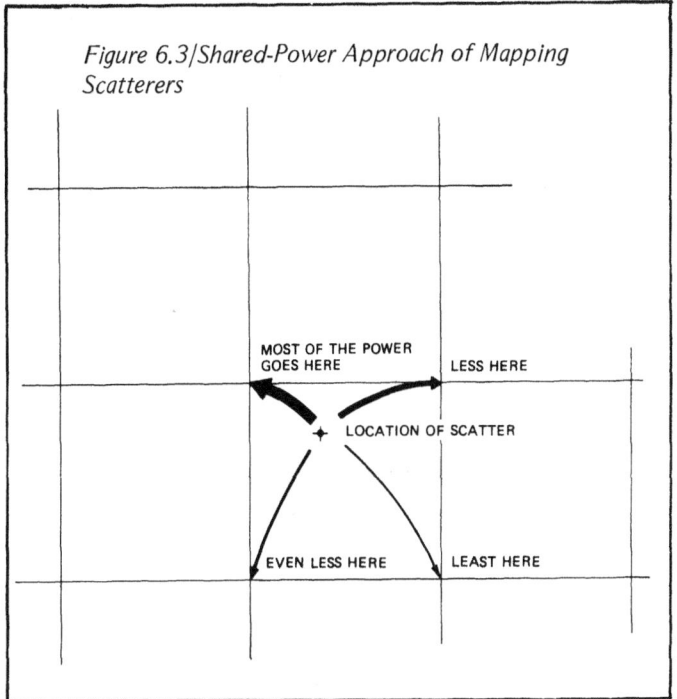

Figure 6.3/Shared-Power Approach of Mapping Scatterers

MOST OF THE POWER GOES HERE   LESS HERE

LOCATION OF SCATTER

EVEN LESS HERE   LEAST HERE

ample, a radar coordinate (u, v) in two dimensions would get mapped as

$$\left.\begin{array}{l} \bar{\sigma}_{m,n} = (1-\alpha)(1-\beta)\bar{\sigma} \\ \bar{\sigma}_{m+1,n} = \alpha(1-\beta)\bar{\sigma} \\ \bar{\sigma}_{m,n+1} = (1-\alpha)\beta\bar{\sigma} \\ \bar{\sigma}_{m+1,n+1} = \alpha\beta\bar{\sigma} \end{array}\right\}$$ (6.18)

where

m = largest integer less than $[(u-u_o)/\Delta u]$

n = largest integer less than $[(v-v_o)/\Delta v]$

$\alpha = (u-u_o)/\Delta u - m$

$\beta = (v-v_o)/\Delta v - n$

The nearest-sample approach is much faster to execute than the shared-power approach, especially if the number of dimensions is large. In Table 6.1 we show the approximate computation times for the various alternatives; each is referenced to the nearest-sample approach in one dimension. There is an obvious tradeoff to be made between the fast nearest-sample approach and the more accurate, but slower, shared-power approach. In the next section, we will discuss the accuracies associated with each approach and postulate what mapping procedure is optimum in a given situation.

Table 6.1/Relative Execution Times for Nearest-Sample and Shared-Power Approaches

	Dimensions			
	1	2	3	4
Nearest-Sample	1	2	3	4
Shared-Power	2	6	14	30

6.4/Sample Spacing

So far, we have stated that the matrix cell $\Delta\tau$ by $\Delta\nu$ must be smaller than a resolution cell. The smaller we make the cell (and consequently the higher the sampling rate), the more accurately we can approximate the continuous functions with discrete ones. However, we always pay the price of increased computation cost for the high sampling rates. If cost were no objective, we could choose unnecessarily high sampling rates. But in most situations, we must make a compromise between accuracy and cost.

Several factors influence the choice of an optimum sampling criterion. First of all, we seldom know the target and clutter characteristics very accurately; therefore, we can tolerate some error due to sampling. Secondly, since the scattering models will usually contain some statistical features, sampling errors will have the same appearance as statistical fluctuations. For example, a

given random sample might fluctuate with a standard deviation of power comparable to the mean. A relative sampling error of a few per cent will be negligible by comparison. Finally, although the sampling error associated with a given scatterer may be high, many such scatterers that are randomly located will tend to reduce the overall net error.

To measure the sampling error, we will relate the two mapping procedures to the effect they have on the receiver response or the two-way antenna power gain pattern (we will simply use the term response for either case). The nearest-sample approach amounts to approximating the power response by rectangular steps, as we show in Fig. 6.4, while the shared-power approach results in linear segments between samples. It is evident that the second method will result in far less peak sampling error than the first if the sampling increment is the same in both cases. But it is interesting to note that the integrated sampling error (if scatterers are uniformly distributed everywhere) is nearly equal for both methods.

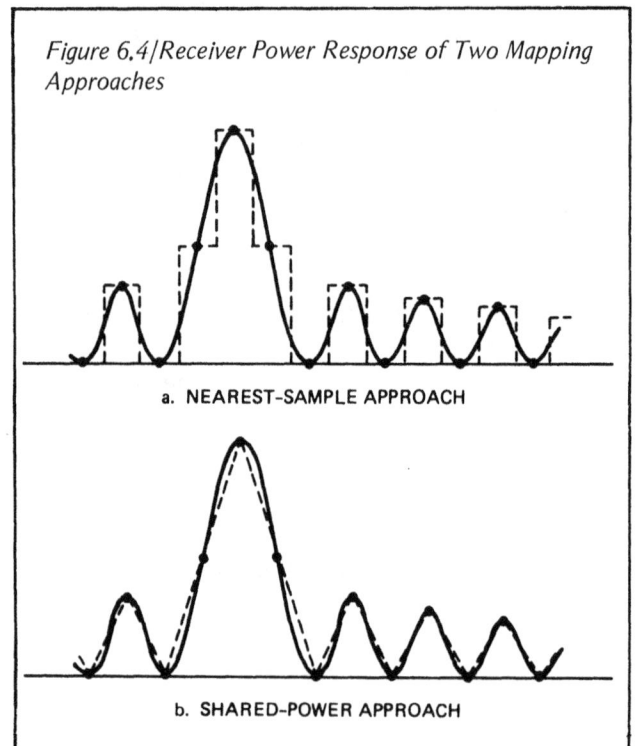

Figure 6.4/Receiver Power Response of Two Mapping Approaches

a. NEAREST-SAMPLE APPROACH

b. SHARED-POWER APPROACH

To investigate the sampling errors, let us distinguish between mainlobe errors (where targets dominate) and sidelobe errors (where clutter dominates). For convenience, we will approximate the mainlobe power response by an inverted parabola between the half-power points as

$$P(u) = 1 - 2(u/u_{3dB})^2 \qquad (6.19)$$

where u is an arbitrary coordinate in radar space and u_{3dB} is the two-sided half-power width. Note that $P(u_{3dB}/2) = 0.5$. We will restrict the mainlobe approximation to $|u| \leqslant u_{3dB}/2$, since a scatterer that is mapped outside of this interval will fall within the mainlobe of another filter, range gate, or antenna beam. For the sidelobe region, we will approximate the normalized power response of one sidelobe by a cosine-squared function as

$$p(u) = \cos^2(\pi u/u_{3dB}) \qquad (6.20)$$

where the null-to-null width is fixed at u_{3dB} (representing a typical sidelobe).

In Table 6.2, we tabulate the mean and rms sampling errors for both approaches in the mainlobe and sidelobe regions, and have normalized the sampling increment Δu to the half-power width of the response as $\Delta = \Delta u/u_{3dB}$. Since the mean and rms errors are comparable in the mainlobe region, especially for large sampling increments, it is desirable to make the mainlobe errors unbiased by multiplying the target power (RCS) by (1 – mean error). As long as $\Delta \leqslant 0.5$, the mean error is given by $-\Delta^2/3$, regardless of the approach. The unbiased rms errors are also tabulated in Table 6.2 for the mainlobe.

For most applications, the simulation output will be sampled at the rate of one sample per resolution cell. In Table 6.2, we have also included this sample spacing for comparison. Although such a sample spacing is marginally ac-

Table 6.2/Response Sampling Errors

Mainlobe Errors are Relative to Mainlobe Peak
Sidelobe Errors are Relative to Sidelobe Peak

$\Delta = \Delta u/u_{3dB}$

		$\Delta=1$	$\Delta=1/2$	$\Delta=1/4$	$\Delta=1/8$	$\Delta=1/16$
Nearest-Sample Approach						
Mainlobe —	mean	.167	-.083	-.021	-.005	-.001
	rms	.223	.171	.084	.042	.021
unbiased rms		.149	.149	.082	.042	.021
Sidelobes —	mean	*	0	0	0	0
	rms	*	.238	.158	.080	.040
Shared-Power Approach						
Mainlobe —	mean	-.083	-.083	-.021	-.0052	-.0013
	rms	.091	.091	.023	.0057	.0014
unbiased rms		.037	.037	.009	.0023	.0006
Sidelobes —	mean	*	0	0	0	0
	rms	*	.075	.075	.0196	.0049

*With $\Delta=1$ the sampling rate is insufficient for the sidelobe region.

curate for targets in the mainlobe, it is grossly inaccurate for clutter in the sidelobe region since we would get on the average only one sample per sidelobe (the Nyquist rate is two samples per cycle). With only one sample per sidelobe, several consecutive samples could fall at response nulls. Thus even though clutter might be present in this region, it would not get included in the integrated response. However, the value of $\Delta = 1$ can be used in certain situations where the accuracy requirements are low, the computation speed requirements are high, and there is no clutter. The advantage of using $\Delta = 1$ is that the convolutions in Sections 6.1 and 6.2 do not have to be performed. Its use will generally be limited to large-scale functional simulations and some real-time simulations, as well as for the convolution with one type of waveform, as we will discuss later.

If the simulation output is sampled at the rate of one sample per resolution cell, the sampling may have to be performed at a higher internal rate to better approximate the real world which is continuous. This situation is generally the case if clutter is present. In some simulation applications, namely in those of tracking and parameter estimation functions, the output sampling rate will be higher than one sample per resolution cell. For such applications, the internal sampling rate can be identical to the output rate, provided the accuracy requirements are met. In any case, the ratio of the sample spacing on output to the internal sample spacing must always be an integer in order for the output samples to be a subset of the internal samples.

In Table 6.2, we note that the relative sampling errors in the sidelobe region are larger than the corresponding errors in the mainlobe region. But, since clutter is the dominant effect in the sidelobe region, we can tolerate relatively large sampling errors on an individual scatterer as long as the integrated sampling error over several scatterers is small. For distributed clutter, this situation is, in fact, the case, since the clutter is reasonably homogeneous over several sidelobes;

it is also the situation for point clutter whenever there are several scatterers and the location of each scatterer in the response is random. The integrated rms sampling error will be reduced by about $1/\sqrt{N}$ for N such scatterers. In general, the mainlobe sampling errors will dominate the accuracy of a simulation.

There is one case where practically no sampling error can be tolerated. This is ground clutter for a ground-based radar (or for any radar on a slowly moving platform). The Doppler frequency of the ground clutter is concentrated at (or near) dc; to reject this strong clutter, we must place a null of the Doppler filter response at dc. In this case, a small sampling error in Doppler would translate into a large error at the filter output.

The only practical solution to this problem is to ensure that there is a sample of the filter response coincident with the Doppler frequency of the intense ground clutter. Otherwise an extremely small value of Δ would have to be used to keep the sampling error less than the filtered residue.

It is not possible to define one procedure that is optimum under all conditions. But in general, all simulation applications fall into three categories. In the first, the computation time is of extreme importance, as with large scale and real-time simulations; the nearest-sample approach is almost mandatory, especially if there are many scatterers to be mapped to radar space. The choice of $\Delta = 1$ or $1/2$ depends on whether clutter is present. The mapping accuracy associated with this category is 15%.

In the second category, the accuracy requirements are higher and the speed requirements are lower. Most simulation applications will fall into this category. Some examples are detection performance evaluation, mapping and display simulation, and some discrimination simulations. Here a sampling error of a few percent will have a negligible effect on the simulation results, in comparison to other factors such as fluctuating targets and poorly defined system losses. In

Table 6.2, there are two choices that satisfy this accuracy requirement — the nearest-sample approach with $\Delta = 1/8$ and the shared-power approach with $\Delta = 1/2$, both resulting in an error of about 4%. The choice between the two depends on whether the mapping of scatterers to radar space, as in Eq. (6.4) or Eq. (6.13), dominates the execution time or whether the convolutions in Eq. (6.2), Eq. (6.3), Eq. (6.10), Eq. (6.11), or Eq. (6.14) dominate. We would use the nearest-sample approach with $\Delta = 1/8$ in the former situation and the shared-power approach with $\Delta = 1/2$ in the latter (the time to perform a discrete convolution is roughly inversely proportional to Δ).

In the third category, accuracy is at a premium, such as in the simulation of tracking and parameter estimation functions. Here the shared-power approach is desirable because we get the highest accuracy with a given sample spacing. In general, a sample spacing (internal to the simulation) that corresponds to the one used in the tracking or parameter estimation algorithm is adequate.

There is one situation where the above comments do not strictly apply; that is, where phase codes are used either on a short pulse (Class A1 waveform) or as the pulse modulation function in a coherent pulse train (Class C waveform). To illustrate this exception, we will give the example of the *Barker* binary phase code. In Fig. 6.5 we show the autocorrelation function of a 5-bit Barker code with the phase sequence $\{0, 0, 0, \pi, 0\}$. The uniform peak sidelobe level is characteristic of the Barker code. The so-called bit length is designated as T_B and it is also the output sample spacing. The pulse length is $T_P = 5T_B$. Note also that the autocorrelation function is uniquely defined by the samples at integer multiples of T_B. Straight-line segments connect adjacent samples. In contrast to most all other types of waveforms, each sidelobe extends over two output samples instead of one. Thus the sampling rate of $\Delta\tau = T_B$ is sufficient in the sidelobe region for the integration of clutter. Since the power response is $|x(\tau)|^2$, it will no longer be made up of straight line segments; however, it will still have a different appearance than the example in Fig. 6.4. The response sampling errors are comparable to those in Table 6.2, except, of course, for the sidelobes with $\Delta = 1$. There is a mean error of $\Delta^2/6$ relative to the mainlobe or sidelobe peak (and for either the nearest-sample or shared-power approach) that can be removed by reducing the RCS of all scatterers by the multiplicative factor $(1 - \Delta^2/3)$. Thus for $\Delta = 1$, we should multiply the RCS of each scatterer by 2/3 to remove the bias. Non-Barker binary phase codes and most polyphase codes are very similar to the Barker codes in their sidelobe behavior, except for an occasional sidelobe that is only as wide as T_B. For most simulation problems, we could treat them in the same manner as the Barker binary phase codes.

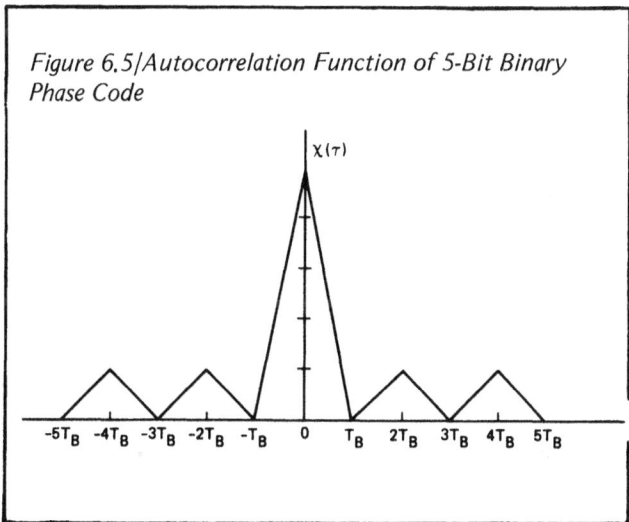

Figure 6.5/Autocorrelation Function of 5-Bit Binary Phase Code

6.5/ Transforming Targets for Ground-Based Radars

Targets expressed in physical space must be transformed to radar space. The transformation is straightforward whenever both the target and radar are in rectangular coordinates. Let us refer to Fig. 6.6 where the radar is located at the origin of the rectangular coordinate system. The x-y plane is horizontal and the azimuth angle θ is measured from the y-axis. The elevation angle ϵ is measured from the x-y plane.

If the target is located at (x, y, z) and moving with the components $(\dot{x}, \dot{y}, \dot{z})$, the range is given by

$$r = (x^2+y^2+z^2)^{1/2} \tag{6.21}$$

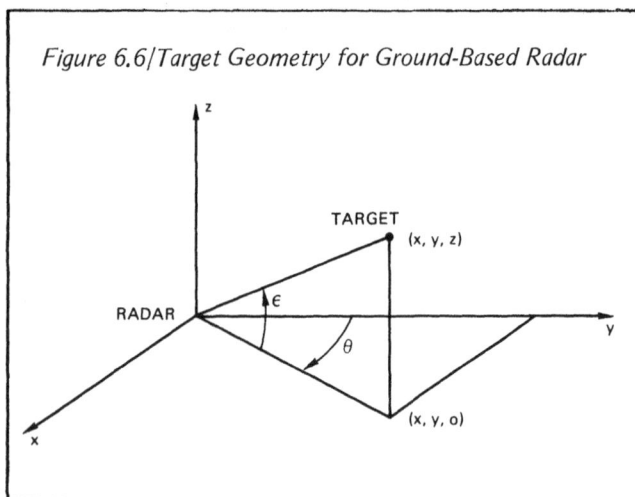

Figure 6.6/Target Geometry for Ground-Based Radar

When we differentiate, we can solve for $\dot{r}$ as

$$\dot{r} = (x\dot{x} + y\dot{y} + z\dot{z})/r \tag{6.22}$$

The azimuth angle, which is necessary to determine the antenna gain from the pattern, is given by

$$\theta = \tan^{-1}(x/y) \tag{6.23}$$

And the elevation angle is given by

$$\epsilon = \sin^{-1}(z/r) \tag{6.24}$$

6.6/ Transforming Ground Clutter for Ground-Based Radars

If the ground clutter model is based on hypothetical terrain, we will begin with the clutter backscatter coefficient, σ_o, expressed as a function of either rectangular (x, y) or polar (r, θ) coordinates. If the former system is used, the ground clutter map must be transformed to (r, θ), since the radar coordinate system is polar. There are two ways to implement this transformation. The first, which we designate as the forward direction, is to transform a given clutter patch of area ΔA and backscatter coefficient σ_o to radar coordinates. The RCS in radar coordinates is $\sigma_o \Delta A$. If the location of the clutter patch is (x, y) in rectangular coordinates, the range and azimuth (r, θ) as defined in Fig. 6.7, are given by

$$r = (x^2+y^2)^{1/2} \tag{6.25}$$

$$\theta = \tan^{-1}(x/y) \qquad\qquad (6.26)$$

Ground clutter will be essentially stationary to a ground-based radar, so we will not have to perform a transformation to Doppler. There will be some random fluctuations caused by the wind (internal motions) which are best modeled statistically in radar space.

In the reverse direction, we begin with a particular cell in radar space at (r, θ) of size $(\Delta r, \Delta\theta)$ and we wish to find the intersection of this cell on the ground. The (x, y) location is just the inverse of Eqs. (6.25) and (6.26) as

$$x = r \sin\theta \qquad\qquad (6.27)$$

$$y = r \cos\theta \qquad\qquad (6.28)$$

The area of the cell intersecting the ground is $\Delta A = r\Delta r\Delta\theta$ and the RCS that gets mapped into that cell is again $\sigma_o\Delta A$.

Both methods of mapping ground clutter to radar space are not equally satisfactory under all conditions. The choice depends on the relationship between the size of the clutter patch and that of the radar cell. If the clutter patch over which σ_o is constant is much larger than the radar cell, then the reverse direction is best since the forward would result in too many gaps in the transformed clutter distribution. However, if the clutter patch is small compared to the radar cell, then the forward direction is best; otherwise detail on the physical map would be lost. We illustrate these relationships in Fig. 6.8.

None of the transformations in Eqs. (6.25) through (6.28) is particularly well suited for rapid computation because each one requires a time-consuming computation (sine, cosine, square-root, or arc-tangent). If these expressions were used in a simulation involving many transformations, most of the time would be spent taking sines, cosines, etc. To be efficient, we must derive algorithms to transform coordinates that do not require the use of the library functions for each point being transformed. In Section 6.9, we show that it is possible to derive

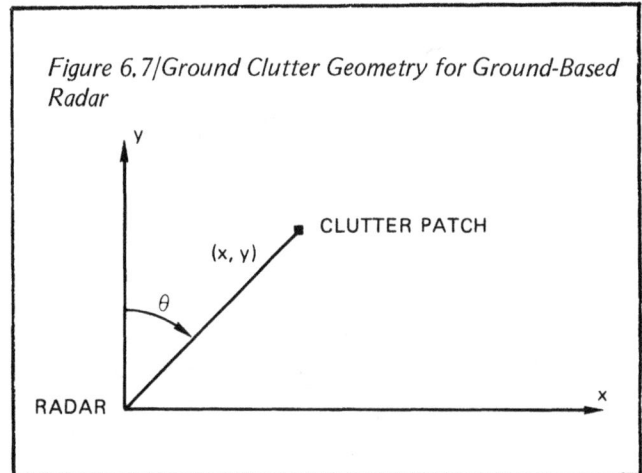

Figure 6.7/Ground Clutter Geometry for Ground-Based Radar

these algorithms when the coordinates are based on a uniformly gridded structure and we proceed along the grid in a regular manner.

When one transforms real terrain to radar space, it is common practice to make the RCS a function of the local terrain slope and to include shadowing. Both effects are easily simulated if we first transform the terrain elevation map in rectangular coordinates to polar coordinates. To perform this transformation, it is best to do it in the reverse direction; that is, to begin with a particular polar-coordinate sample (r, θ) and find what terrain elevation corresponds to that sample. To do the transformation in the forward direction would result in some polar-coordinate cells containing several terrain points and other cells containing none (such an effect is satisfactory when transforming RCS, but not desirable when transforming terrain elevation). Interpolation could be used to fill in the vacant cells, however.

The next step is to compute RCS. We could do this on the basis of the local 3-D terrain slope, but it is much simpler (and probably as realistic) just to assume that the RCS is a function of the 2-D profile of elevation vs. range at a constant value of θ. The final step is to compute which areas are shadowed by terrain features in the

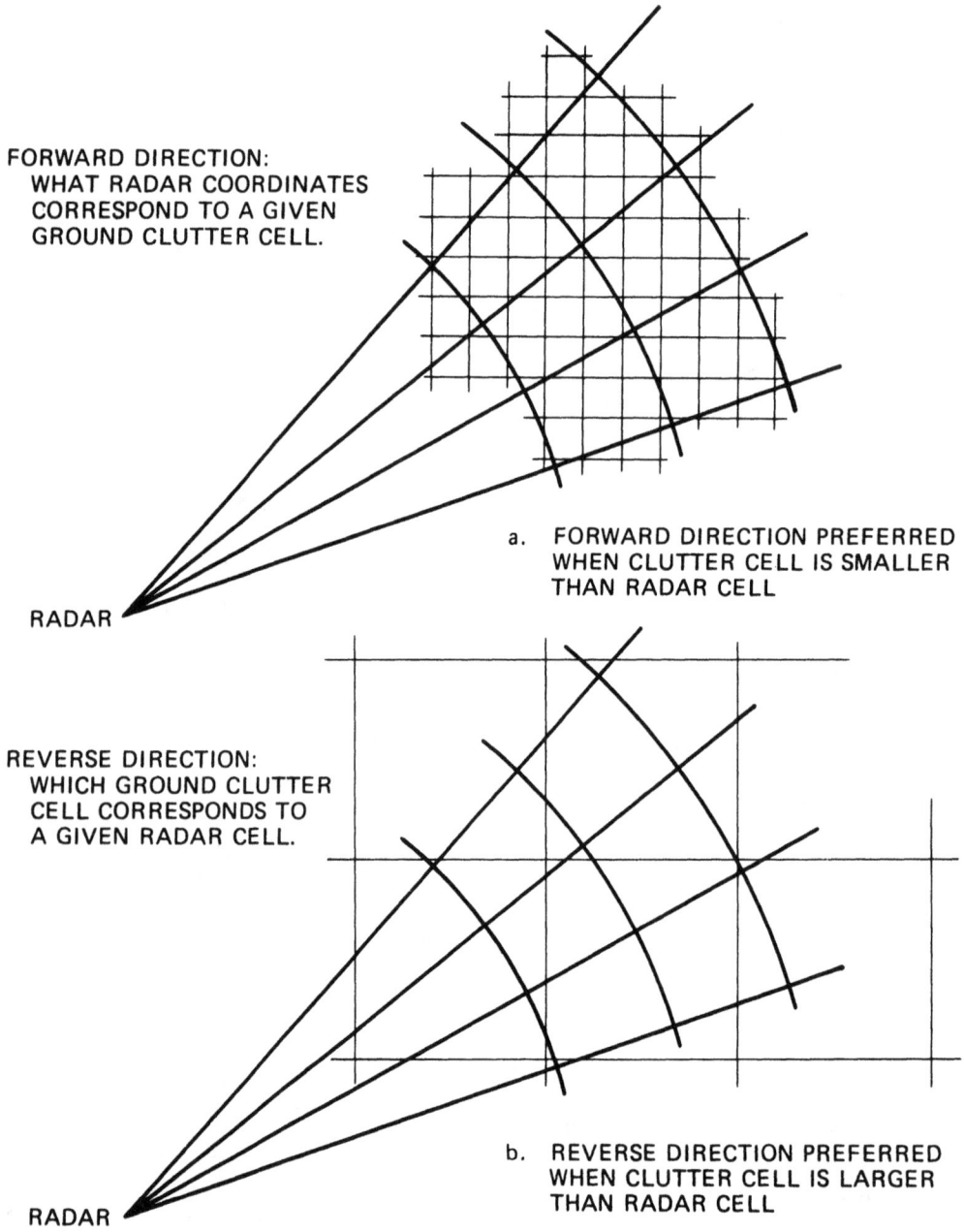

Figure 6.8/Two Methods of Mapping Ground Clutter
to Radar Space for a Ground-Based Radar

FORWARD DIRECTION:
  WHAT RADAR COORDINATES
  CORRESPOND TO A GIVEN
  GROUND CLUTTER CELL.

RADAR

a.   FORWARD DIRECTION PREFERRED
      WHEN CLUTTER CELL IS SMALLER
      THAN RADAR CELL

REVERSE DIRECTION:
  WHICH GROUND CLUTTER
  CELL CORRESPONDS TO
  A GIVEN RADAR CELL.

RADAR

b.   REVERSE DIRECTION PREFERRED
      WHEN CLUTTER CELL IS LARGER
      THAN RADAR CELL

foreground. This computation is easily accomplished by tracing rays from the radar to each sample on the 2-D elevation profile. Beginning at the near range, if the slope of a given ray is less than the largest previous slope, the terrain sample corresponding to that ray is shadowed (the RCS is set to zero). There is more discussion on this subject in Section 10.1.

6.7/Transforming Targets for Airborne Radars

The transformation for targets is not so straightforward if the radar is located on an airborne platform. The reason is that the antenna coordinate system is not, in general, compatible with the physical space geometry. The simplest case is for level flight. Let us refer to Fig. 6.9, which is identical to Fig. 6.6, except that the radar is located at the coordinate (x_R, y_R, z_R) and the target at (x_T, y_T, z_T). We lose no generality if we assume that the radar platform velocity vector is parallel to the y-axis. Again let the components of the target velocity be $(\dot{x}_T, \dot{y}_T, \dot{z}_T)$. The range is now given by

$$r = [(x_T - x_R)^2 + (y_T - y_R)^2 + (z_T - z_R)^2]^{1/2} \quad (6.29)$$

When we differentiate, we can again solve for $\dot{r}$ as

$$\dot{r} = \frac{(x_T - x_R)(\dot{x}_T - \dot{x}_R) + (y_T - y_R)(\dot{y}_T - \dot{y}_R) + (z_T - z_R)(\dot{z}_T - \dot{z}_R)}{r} \quad (6.30)$$

where $\dot{x}_R = \dot{z}_R = 0$ and $\dot{y}_R = V$, the platform velocity. Similarly,

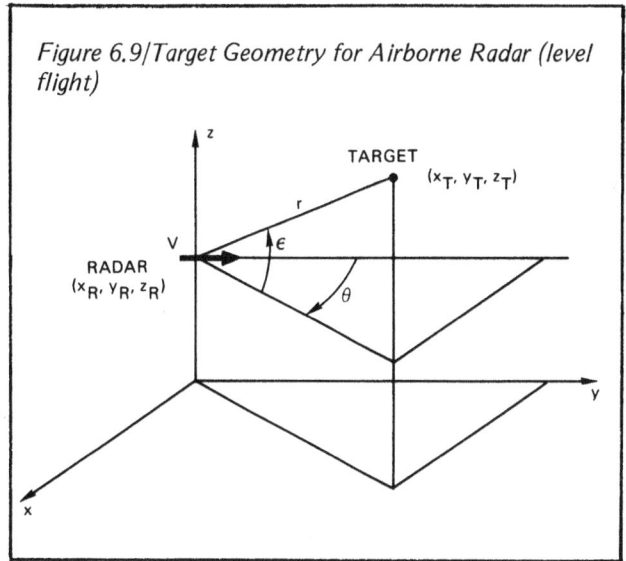

Figure 6.9/Target Geometry for Airborne Radar (level flight)

$$\theta = \tan^{-1}[(x_T - x_R)/(y_T - y_R)] \quad (6.31)$$

$$\epsilon = \sin^{-1}[(z_T - z_R)/r] \quad (6.32)$$

The above transformations to (θ, ϵ) are useful only if that is the antenna coordinate system; in many cases, it will not be. We will not go into the various possibilities here.

In the above transformation we assumed level flight. The situation for nonlevel flight is more complex. Let the flight vector be located in the y-z plane at a dive angle δ below the horizontal, as we show in Fig. 6.10. The velocity components of the flight vector are $(0, V\cos\delta, -V\sin\delta)$, which we can substitute directly into Eq. (6.30). For the purposes of our discussion, let us assume that the antenna coordinate system is azimuth-elevation (θ', ϵ') measured in the plane containing the flight vector and the x-axis (i.e., the roll angle is zero). After some manipulation, we can derive the transformation from (θ, ϵ) given by Eqs. (6.31) and (6.32) to (θ', ϵ') as

$$\tan\theta' = \frac{\cos\epsilon \sin\theta}{\cos\epsilon \cos\theta \cos\delta - \sin\epsilon \sin\delta} \quad (6.33)$$

$$\sin\epsilon' = \cos\epsilon \cos\theta \sin\delta + \sin\epsilon \cos\delta \quad (6.34)$$

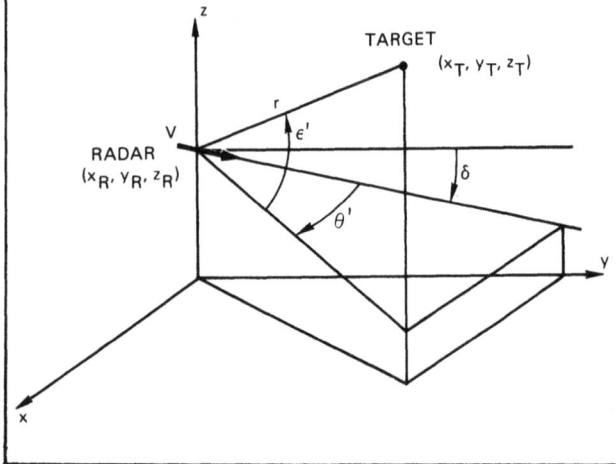

Figure 6.10/Target Geometry for Airborne Radar (nonlevel flight)

Again, this is useful only if the antenna coordinate system is (θ', ϵ').

6.8/Transforming Ground Clutter for Airborne Radars

The mapping or transformation of ground clutter to radar space is complicated when the radar is on an airborne platform. Each point on the ground will get mapped into a unique coordinate in 4-D radar space — range, Doppler, and two antenna angles. For level flight, certain simplifications are possible; for nonlevel flight, the most

difficult problem is to compute the antenna coordinates of a point on the ground.

It is practical to perform the mapping only in the forward direction for airborne radars. A point on the ground must be mapped into radar coordinates. If we try to do the mapping in the reverse direction where we determine the intersection of a range-Doppler cell on the ground, we will have difficulty in computing the area of this intersection. The area will be a complex function of both range and azimuth; moreover, certain special cases are sometimes encountered. We avoid these difficulties by mapping in the forward direction.

If the ground clutter model is based on hypothetical terrain, the clutter backscatter coefficient σ_o will be expressed as a function of either rectangular (x, y) or polar (ρ, θ) coordinates (in the latter case, ρ is ground range). Let us refer to Fig. 6.11, where the ground is located in the x-y plane and the radar on the z-axis at an altitude h above the ground plane (we lose little generality if we assume a flat earth, as we will discuss later). We will constrain the platform velocity vector to lie within the y-z plane at a dive angle δ relative to the horizontal. For any point

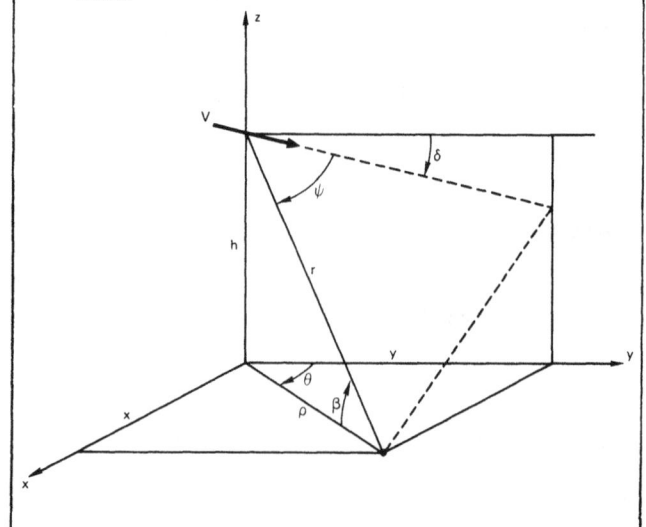

Figure 6.11/Ground Clutter Geometry for Airborne Radar

on the ground at a coordinate (x, y) the slant range is given by

$$r = (x^2 + y^2 + h^2)^{1/2} \qquad (6.35)$$

From Eq. (6.30), we can show that the range rate is given by

$$\dot{r} = -V[(y/r)\cos\delta + (h/r)\sin\delta] \qquad (6.36)$$

where V is the platform velocity. The azimuth angle referenced to the (x, y, z) coordinate system is given by

$$\theta = \tan^{-1}(x/y) \qquad (6.37)$$

and the grazing angle by

$$\beta = \sin^{-1}(h/r) \qquad (6.38)$$

If the flight is level ($\delta = 0$), the depression angle measured from the radar to the point is also β and the elevation angle (measured at the radar) is

$$\epsilon = -\beta \qquad (6.39)$$

If the flight is not level, we must transform (θ, ϵ) to the antenna coordinate system. If that coordinate system is also azimuth-elevation designated by (θ', ϵ'), then we can apply the transformations in Eqs. (6.33) and (6.34).

The expressions in Eqs. (6.35) and (6.37) are not computationally efficient if there are many points to be transformed, as is usually the case with ground clutter. However, we note that these expressions are essentially the same as those in Eqs. (6.25) and (6.26). We also derive efficient algorithms in Section 6.9 for the airborne geometry. The expression for range rate in Eq. (6.36) is computationally efficient since it can be reduced to one addition and one division if we increment y while holding x constant, or to just one division if we increment x while holding y constant. The remaining expression in Eq. (6.38) will not usually be evaluated in that form; instead, we will most likely need $\cos\epsilon = -h/r$ in order to compute the transformation to antenna coordinates in Eqs. (6.33) and (6.34). The fast algorithm for the square-root in Chapter 9 can be used here.

The transformation from rectangular coordinates to radar coordinates is somewhat cumbersome to implement in many cases, especially if the range interval of interest is narrow. As we show in Fig. 6.12, the ground map will have to be large enough to include the maximum range; much of the map within the nearest range, though, will be unused. A simpler arrangement would be to express the ground map in polar (ρ, θ) coordinates. Now the slant range is given by

$$r = (\rho^2 + h^2)^{1/2} \qquad (6.40)$$

Since $x/\rho = \sin\theta$, $y/\rho = \cos\theta$, and $\rho/r = \cos\beta$, we can write

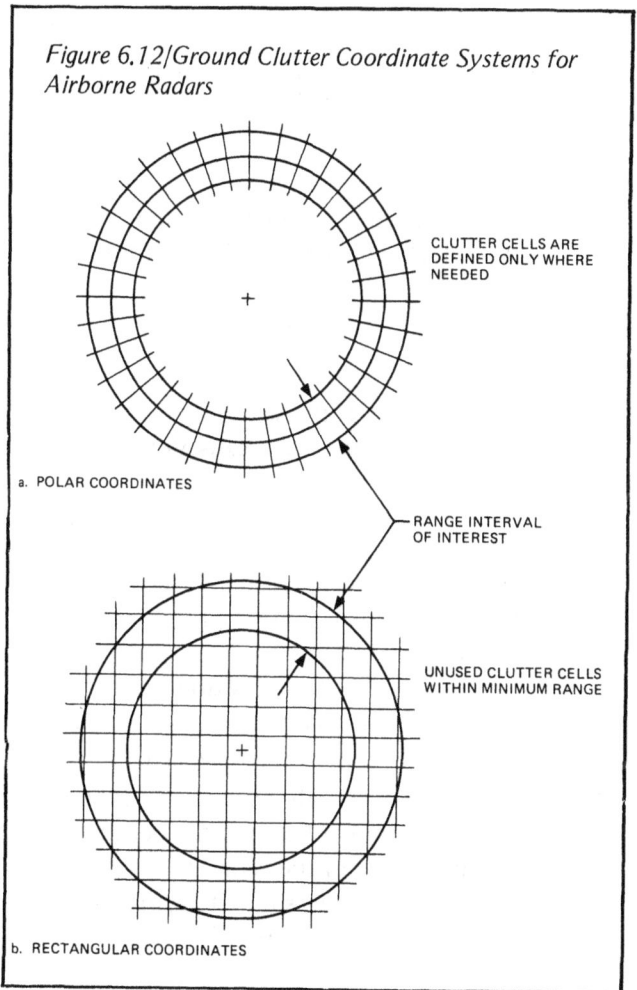

Figure 6.12/Ground Clutter Coordinate Systems for Airborne Radars

CLUTTER CELLS ARE DEFINED ONLY WHERE NEEDED

a. POLAR COORDINATES

RANGE INTERVAL OF INTEREST

UNUSED CLUTTER CELLS WITHIN MINIMUM RANGE

b. RECTANGULAR COORDINATES

$$\frac{y}{r} = \frac{y}{\rho} \cdot \frac{\rho}{r} = \cos\theta \, \cos\beta \qquad (6.41)$$

In Eq. (6.38), $h/r = \sin\beta$ so that we can write Eq. (6.36) as

$$\dot{r} = -V[\cos\theta \, \cos\beta \, \cos\delta + \sin\beta \, \sin\delta] \qquad (6.42)$$

The transformation from polar coordinates to radar coordinates is easily evaluated if we increment θ while we keep ρ fixed. In this case r and β are constant. Furthermore if $V\cos\theta$ has been precomputed in a table, then Eq. (6.42) reduces to one multiplication and one addition. The choice of the sampling increment $\Delta\theta$ is straight-forward. Let us first write the Doppler frequency ($\nu = -2\dot{r}/\lambda$) in terms of the angle ψ measured from the flight vector as

$$\nu = (2V/\lambda) \cos\psi \qquad (6.43)$$

A change in ψ by an amount $\Delta\psi$ will change the Doppler frequency by

$$|\Delta\nu| = (2V/\lambda)\Delta\psi|\sin\psi| \qquad (6.44)$$

The largest change in Doppler frequency for a given $\Delta\psi$ occurs when $\psi = \pi/2$ (broadside). Neglecting the difference between $\Delta\psi$ and $\Delta\theta$, we can write

$$\Delta\nu \geqslant (2V/\lambda)\Delta\theta \qquad (6.45)$$

If we designate $\Delta\nu$ as the sampling increment in Doppler, then a value of $\Delta\theta$ satisfying

$$\Delta\theta \leqslant \frac{\lambda\Delta\nu}{2V} \qquad (6.46)$$

will ensure that no cell in radar space will be skipped in the transformation process. Usually it is only necessary to satisfy the equality in Eq. (6.46). Of course, if the resolution in azimuth on the basis of the antenna pattern results in a smaller value of $\Delta\theta$, we must use that value.

Real terrain is not usually used as the basis for ground clutter in an airborne radar because of the dynamic nature of the radar platform. It is generally satisfactory to simulate typical or representative ground clutter that is based on statistical models. If we were to use real terrain, we

would have to implement a procedure similar to that discussed in Section 6.6. Real terrain is used in the simulation of airborne mapping radars; but in this case, we will treat the terrain and terrain features as the target, not as clutter. For airborne mapping radars, the coordinate transformations can usually be simplified much more than in other radar simulations (e.g., level flight and small depression angles can often be assumed).

Before we leave the subject of airborne radars, let us discuss the effect of the spherical earth. In Fig. 6.13, we have drawn one ray from the radar to the spherical earth at a slant range r and another ray to the tangent plane representing the flat earth, also at a slant range r. The first observation we can make is that the grazing angle for a spherical earth is slightly smaller than the one for a plane earth. If we let β_s denote the grazing angle for a spherical earth, we can derive

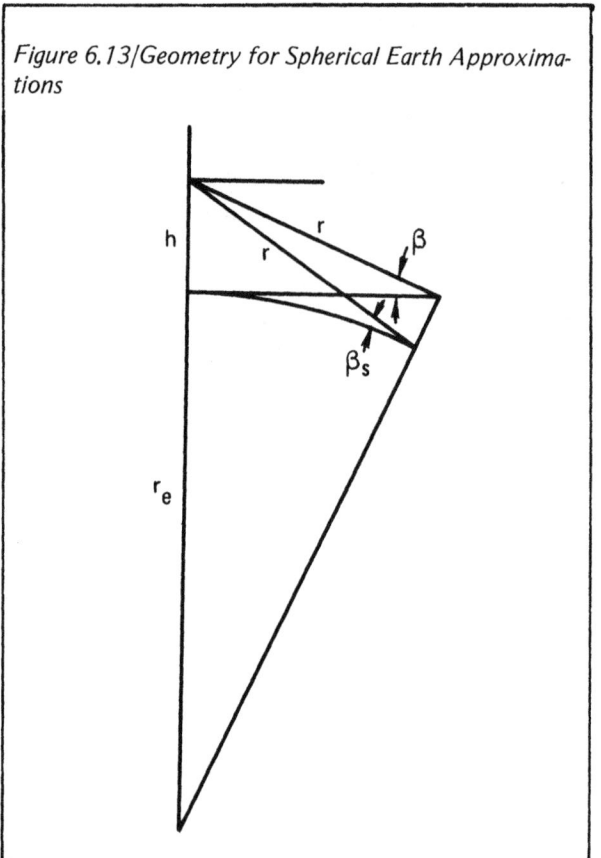

Figure 6.13/Geometry for Spherical Earth Approximations

$$\sin \beta_s = \frac{h^2 + 2r_e h - r^2}{2rr_e} \qquad (6.47)$$

where r_e is the radius of the earth for propagation purposes. A very accurate approximation, especially for the shallower grazing angles, is obtained by neglecting h^2 in Eq. (6.47), which results in

$$\sin \beta_s = h/r - r/2r_e \qquad (6.48)$$

where the error is bounded by $h/2r_e$. For all airborne radars, Eq. (6.48) is accurate to at least three decimal places (the approximation begins to break down for radars on board spacecraft). Note that $h/r = \sin\beta$, where β is the grazing angle for a flat earth. Just as $\sin\beta$ is reduced by $r/2r_e$ in Eq. (6.48), the sine of the depression angle measured at the radar is increased by the same factor over the flat earth value (i.e., $h/r+r/2r_e$). We can also use Eq. (6.48) to compute the range to the horizon by setting $\beta_s = 0$ and obtaining

$$r_H = \sqrt{2r_e h} \qquad (6.49)$$

The difference between the ground ranges measured along the flat earth and spherical earth is approximately

$$\Delta\rho = rh/2r_e \qquad (6.50)$$

At the high altitude of $h = 15$ km and the maximum range of $r_H = 500$ km, the difference in ground ranges amounts to only 440 meters. Such a difference is negligible in practically all simulation applications. It just means that a slightly different point on the ground gets mapped into a given radar cell. The most important consequence of the spherical earth is the correction for the grazing angle in Eq. (6.48), which is trivial to implement.

6.9/Fast Transformation Algorithms

As we mentioned in Section 6.6, the equations used to transform a point in rectangular coordinates to polar coordinates, or vice versa, are not particularly fast to compute. However, if the coordinate systems are based on a uniformly gridded structure, it is possible to develop algorithms that are computationally fast. In going from polar to rectangular coordinates, we will assume that r and θ are incremented by Δr and $\Delta\theta$; in going from rectangular to polar coordinates, x and y will be incremented uniformly by Δx and Δy.

(r, θ) → (x, y) for Constant θ

The simplest expressions to derive are in the reverse direction; therefore, we will discuss them first. If we increment r while holding θ constant in Eqs. (6.27) and (6.28), we do not have to repeat the $\sin\theta$ and $\cos\theta$ calculations for each new value of r. Thus if we use the subscript k to denote the kth sample of x, y, and r, the values at the k+1st sample are

$$r_{k+1} = r_k + \Delta r$$
$$x_{k+1} = s\, r_{k+1} \qquad (6.51)$$
$$y_{k+1} = c\, r_{k+1}$$

where $c = \cos\theta$ and $s = \sin\theta$. Note that the subscripts on x and y do not imply that these values are uniformly incremented. It is trivial, however, to find which rectangular cell contains the coordinate (x_k, y_k).

(r, θ) → (x, y) for Constant r

If we increment θ while holding r constant, the

recursion relationships are slightly more complicated. Let us first write

$$x + \Delta x = r \sin(\theta + \Delta\theta)$$

$$y + \Delta y = r \cos(\theta + \Delta\theta)$$

(6.52)

where Δx and Δy now denote the change in x and y. If we expand the expressions on the right, we can write

$$x_{k+1} = x_k c_\Delta + y_k s_\Delta$$

$$y_{k+1} = y_k c_\Delta - x_k s_\Delta$$

(6.53)

where $c_\Delta = \cos\Delta\theta$ and $s_\Delta = \sin\Delta\theta$. We must use Eqs. (6.27) and (6.28) to begin the iteration.

(x, y) → (r, θ) for Constant x

For the forward direction, it is not possible to derive exact expressions that are any more efficient than Eqs. (6.25) and (6.26). However, we can derive approximations that are both simple and highly accurate. In Eq. (6.25), let us increment y by Δy while holding x constant. Then

$$(r+\Delta r)^2 = x^2 + (y+\Delta y)^2$$

(6.54)

By rearranging terms, we can write

$$\Delta r = \Delta y \, \frac{2y+\Delta y}{2r+\Delta r}$$

(6.55)

We do not yet have an explicit solution for Δr since it appears on both sides of Eq. (6.55). But if we drop Δy and Δr on the right-hand side of Eq. (6.55) in the interest of obtaining a simple expression, we obtain

$$\Delta r = y\Delta y/r$$

(6.56)

We will show the consequence of this approximation later.

To obtain $\Delta\theta$, let us increment y by Δy while holding x constant in Eq. (6.26). After some manipulation, we can write

$$\Delta\theta = \tan^{-1} \left[\frac{-x\Delta y}{r^2 + y\Delta y} \right]$$

(6.57)

which is even more complicated than Eq. (6.26). But if we assume that the argument of the arctangent is small, we obtain

$$\Delta\theta = \frac{-x\Delta y}{r^2 + y\Delta y}$$

(6.58)

Let us go one step further and omit the $y\Delta y$ in the denominator so that we can write

$$\Delta\theta = -x\Delta y/r^2$$

(6.59)

The recursion relations then become

$$r_{k+1} = r_k + y_k\Delta y/r_k$$

$$\theta_{k+1} = \theta_k - x\Delta y/r^2_{k+1}$$

$$y_{k+1} = y_k + \Delta y$$

(6.60)

Note the use of r_{k+1} in the second expression instead of r_k. It has been found empirically that this results in less error than if r_k is used. In the use of Eq. (6.60), we must begin the iteration by using the exact expressions in Eqs. (6.25) and (6.26) to compute the initial coordinate (r_0, θ_0).

The recursion relationships in Eq. (6.60) are stable if we iterate in the direction of increasing range. The errors have been determined empirically to be bounded by

$$|\mathcal{E}_r| \leqslant 0.5 \, |\Delta y \sin^2\theta_0|$$

(6.61)

$$|\mathcal{E}_\theta| \leqslant 0.5 \, |(\Delta y/r_0)\sin\theta_0|$$

(6.62)

where (r_0, θ_0) is the initial coordinate. These error bounds are quite small in practice. We must remember that it is practical to transform from (x, y) to (r, θ) only when the rectangular cell $(\Delta x, \Delta y)$ is smaller than the radar cell $(\Delta r, \Delta\theta)$, otherwise there will be gaps in the transformed function. Since the range error is never larger than $\Delta y/2$ in Eq. (6.61), the range error will be even smaller compared to the range cell Δr. The same argument applies to the error in azimuth angle. But of even more importance than absolute errors is the fact that the relative errors among adjacent cells will be extremely small.

If we iterate in the direction of decreasing range, the recursion relationship for θ is not stable. Whenever the range of the iteration covers both positive and negative values of y, we must begin at the x-axis and proceed separately in each direction away from the x-axis.

$(x, y) \rightarrow (r, \theta)$ for Constant y

We can interchange x and y in the previous derivation if we replace θ by $\pi/2-\theta$. Thus the recursion relationships are given by

$$r_{k+1} = r_k + x_k\Delta x/r_k$$

$$\theta_{k+1} = \theta_k + y\Delta x/r_{k+1}^2 \qquad (6.63)$$

$$x_{k+1} = x_k + \Delta x$$

The error bounds are

$$|\varepsilon_r| \leq 0.5 \,|\Delta x \cos^2\theta_o| \qquad (6.64)$$

$$|\varepsilon_\theta| \leq 0.5 \,|(\Delta x/r_o)\cos\theta_o| \qquad (6.65)$$

Airborne Geometry

In Section 6.8, we indicated that the transformation from rectangular (x, y) to radar coordinates for an airborne geometry is as inefficient to implement as it is for the ground-based radar geometry. The problem is to find the radar coordinate (r, θ) that corresponds to the ground

coordinate (x, y). Since the slant range r in Eq. (6.35) contains an extra h^2 term, we can modify Eq. (6.60) or Eq. (6.63) to take this term into account. The resulting recursion relationships are

$$\rho_{k+1} = \rho_k + y_k\Delta y/\rho_k$$

$$r_{k+1} = r_k + y_k\Delta y/r_k \qquad (6.66)$$

$$\theta_{k+1} = \theta_k - x\Delta y/\rho_{k+1}^2$$

$$y_{k+1} = y_k + \Delta y$$

where we have kept x fixed, or

$$\rho_{k+1} = \rho_k + x_k\Delta x/\rho_k$$

$$r_{k+1} = r_k + x_k\Delta x/r_k \qquad (6.67)$$

$$\theta_{k+1} = \theta_k + y\Delta x/\rho_{k+1}^2$$

$$x_{k+1} = x_k + \Delta x$$

where we have kept y fixed. In either case, we must use Eqs. (6.35), (6.37), and

$$\rho = (x^2 + y^2)^{1/2}$$

to begin the iteration.

Chapter 7

Sampled Signals

In a digital computer implementation all signals must be sampled. In this chapter we will derive Fourier transform relationships for those sampled in the time domain; then we will extend the results to both time and frequency. Definite rules for sampling will be established.

There are numerous references for the implementation of discrete Fourier transforms and fast computational algorithms. The most extensive treatment is the book by Brigham [Ref. 1]; the paper by Bergland [Ref. 2] provides a very readable introduction to the subject. Selected issues of the *IEEE Transactions on Audio and Electroacoustics* [Refs. 3 and 4] are also of interest here, as is the monograph on digital signal processing by the IEEE Press [Ref. 5].

7.1/Sampling in the Time Domain

Let us begin with an arbitrary complex signal (which could also be a complex modulation function) that is continuous and defined by the Fourier transform pair

$$X(f) = \int_{-\infty}^{\infty} x(t)\, e^{-j2\pi ft}\, dt \qquad (7.1)$$

$$x(t) = \int_{-\infty}^{\infty} X(f)\, e^{j2\pi ft}\, df \qquad (7.2)$$

In Table 7.1 we summarize the properties of the continuous Fourier transform pair. We can approximate $X(f)$ by sampling $x(t)$ at discrete times $k\Delta t$ and summing as

Table 7.1/Continuous Fourier Transform Relationships

$$X(f) = \int_{-\infty}^{\infty} x(t)\, e^{-j2\pi ft}\, dt$$

$$x(t) = \int_{-\infty}^{\infty} X(f)\, e^{j2\pi ft}\, df$$

$$E_c = \int_{-\infty}^{\infty} |x(t)|^2\, dt = \int_{-\infty}^{\infty} |X(f)|^2\, df$$

$$\int_{-\infty}^{\infty} e^{j2\pi ft}\, df = \delta(t)$$

$$\int_{-\infty}^{\infty} e^{-j2\pi ft}\, dt = \delta(f)$$

$$X_s(f) = \Delta t \sum_{k=-\infty}^{\infty} x(k\Delta t)\, e^{-j2\pi fk\Delta t} \qquad (7.3)$$

where the subscript s is used to denote the transform of a sampled function. At this point we will assume that the index k is incremented over all values from $-\infty$ to ∞, even though $x(t)$ may be time limited.

A function sampled in the time domain will be repetitive in the frequency domain. This fact is evident if we replace f in Eq. (7.3) by $f+m/\Delta t$ where m is any integer. We obtain

$$X_s(f+m/\Delta t) = \Delta t \sum_{k=-\infty}^{\infty} x(k\Delta t)\, e^{-j2\pi(fk\Delta t+mk)} \qquad (7.4)$$

Since the product mk is still an integer, the phase (modulo 2π) remains unchanged. Thus we can omit the term mk and the result is identical to Eq. (7.3). Hence

$X_s(f+m/\Delta t) = X_s(f)$ (7.5)

Hence, $X_s(f)$ is a periodic function with a *repetition frequency* given by

$f_r = 1/\Delta t$ (7.6)

In Eq. (7.3) we can substitute the Fourier transform in Eq. (7.2) for $x(k\Delta t)$. The result can be rewritten as

$$X_s(f) = \int_{-\infty}^{\infty} X(f') \left[\Delta t \sum_{k=-\infty}^{\infty} e^{-j2\pi(f-f')k\Delta t} \right] df'$$ (7.7)

For the moment, let us concentrate on the factor within the braces. Based on the similarity between this factor and the continuous integral in Eq. (4.28), it is possible to derive the result for sampled signals as a repeated sequence of Dirac-delta functions as

$$\Delta t \sum_{k=-\infty}^{\infty} e^{-j2\pi f k\Delta t} = \sum_{m=-\infty}^{\infty} \delta(f-mf_r)$$ (7.8)

If we use this result in Eq. (7.7), we obtain

$$X_s(f) = \sum_{m=-\infty}^{\infty} X(f-mf_r)$$ (7.9)

which relates the transform of the sampled signal to the transform of the continuous signal. We illustrate this result in Fig. 7.1, where the original spectrum in (a) is repeated at multiples of the repetition frequency in (b).

Since $X_s(f)$ is periodic while the original spectrum, $X(f)$, is not, the two spectra are not comparable for finite values of Δt if we are concerned with the entire frequency axis. However, if we restrict our interest to a finite band of frequencies, we can readily compare the two spectra, provided Δt is properly chosen. The first requirement is that $f_r = 1/\Delta t$ must be larger than the band of interest; the second has to do with minimizing *aliasing* or *foldover*, which is the overlap of the repeated responses of $X_s(f)$. If most or all of the energy of $X(f)$ is confined to

an interval of width B, which we denote as the *bandwidth*, then little or no energy will foldover if

$f_r \geqslant B$ (7.10)

and, by the use of Eq. (7.6), we can write the equivalent inequality

$\Delta t \leqslant 1/B$ (7.11)

If the original spectrum is bandlimited where there is no energy outside the band of width B, then $X_s(f)$ will be identical to $X(f)$ within the repetition interval, provided that Eq. (7.10) is satisfied. The problem in choosing the proper size of Δt is most critical when the signal is not bandlimited. We will give some guidelines on this matter in Section 7.8.

The Fourier transform of $X_s(f)$ is the sequence of time samples $x(k\Delta t)$. Since $X_s(f)$ is repetitive

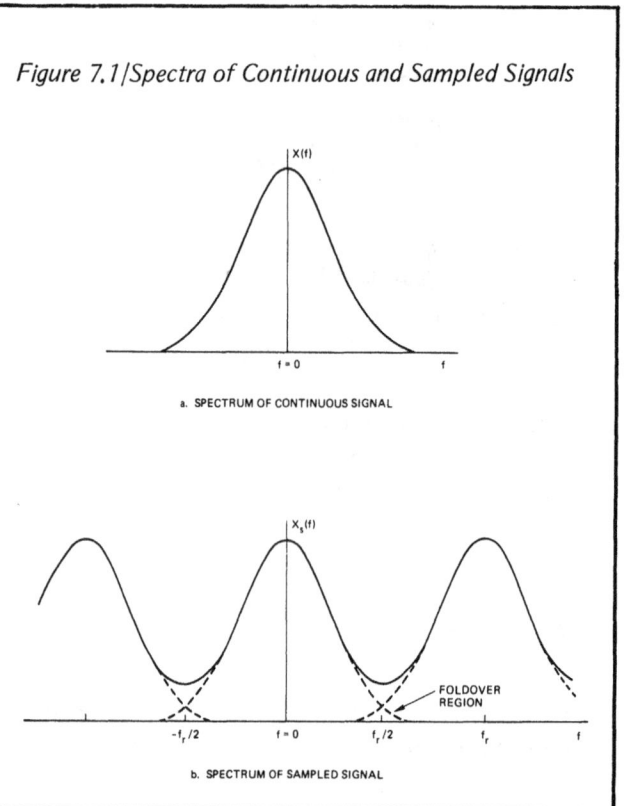

Figure 7.1/Spectra of Continuous and Sampled Signals

a. SPECTRUM OF CONTINUOUS SIGNAL

b. SPECTRUM OF SAMPLED SIGNAL

over the interval f_r, it is of interest to define the time function corresponding to only one period of $X_s(f)$ as

$$x'(t) = \int_0^{f_r} X_s(f) \, e^{j2\pi ft} \, df \qquad (7.12)$$

Let us replace $X_s(f)$ in Eq. (7.12) by Eq. (7.3), where we first change the subscript k to m and interchange the order of summation and integration to obtain

$$x'(t) = \Delta t \sum_{m=-\infty}^{\infty} x(m\Delta t) \int_0^{f_r} e^{j2\pi f(t-m\Delta t)} df \qquad (7.13)$$

Upon integration, we obtain

$$x'(t) = \sum_{m=-\infty}^{\infty} x(m\Delta t) \, \mathrm{sinc}(f_r t - m) e^{j\pi(f_r t - m)} \qquad (7.14)$$

where $\mathrm{sinc}(u) = (\sin\pi u)/\pi u$. In particular, we note that for integral values of $u=k$,

$$\begin{aligned} \mathrm{sinc}(k) &= 1, \quad k = 0 \\ &= 0, \quad \text{otherwise} \end{aligned} \qquad (7.15)$$

Thus at the samples $t = k\Delta t$, only the $m = k$ term in Eq. (7.14) is nonzero so that

$$x'(k\Delta t) = x(k\Delta t) = \int_0^{f_r} X_s(f) \, e^{j2\pi fk\Delta t} \, df \qquad (7.16)$$

In between the samples $x'(t)$ is not coincident with $x(t)$. We illustrate this phenomenon in Fig. 7.2. The expressions in Eqs. (7.3) and (7.16) for a Fourier transform pair.

We can derive the energy relationships corresponding to Eqs. (7.3) and (7.16) by first writing the energy in the time domain for sampled signals as

$$E_c = \Delta t \sum_{k=-\infty}^{\infty} |x(k\Delta t)|^2 \qquad (7.17)$$

Now we substitute Eq. (7.16) for $x(k\Delta t)$ and interchange the order of integration and summation to obtain

$$E_c = \int_0^{f_r}\int X_s(f_1) X_s{}^*(f_2) \left\{ \Delta t \sum_{k=-\infty}^{\infty} e^{j2\pi(f_1-f_2)k\Delta t} \right\} df_1 df_2 \qquad (7.18)$$

The term within the braces is given by Eq. (7.8). But since the integration in Eq. (7.18) is over only one frequency repetition interval, only the $m = 0$ term in Eq. (7.8) is used and the integrand is non-zero only for $f_1 = f_2$. Hence

$$E_c = \int_0^{f_r} |X_s(f)|^2 \, df \qquad (7.19)$$

Since the Δt factor in Eqs. (7.3) and (7.17) acts only as a scale factor, it is sometimes convenient to combine it with the time sample as

$$\hat{x}(k\Delta t) = \Delta t \cdot x(k\Delta t) \qquad (7.20)$$

where the caret ^ signifies the rescaled variable. Eq. (7.3) then can be written as

$$X_s(f) = \sum_{k=-\infty}^{\infty} \hat{x}(k\Delta t) \, e^{-j2\pi fk\Delta t} \qquad (7.21)$$

If we make use of Eq. (7.6), we can rewrite Eq. (7.16) as

$$\hat{x}(k\Delta t) = \frac{1}{f_r} \int_0^{f_r} X_s(f) \, e^{j2\pi fk\Delta t} \, df \qquad (7.22)$$

These two expressions constitute a Fourier transform pair. Eqs. (7.17) and (7.19) can be rewritten as

$$\Delta t \cdot E_c = \sum_{k=-\infty}^{\infty} |\hat{x}(k\Delta t)|^2 = \frac{1}{f_r} \int_0^{f_r} |X_s(f)|^2 \, df \qquad (7.23)$$

We summarize the results for sampling signals in the time domain in Table 7.2.

Figure 7.2/Fourier Transform Relationships for Sampled Signal

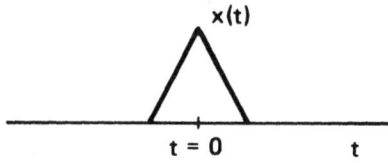

HERE IS THE ORIGINAL FUNCTION . . . .   AND ITS SPECTRUM

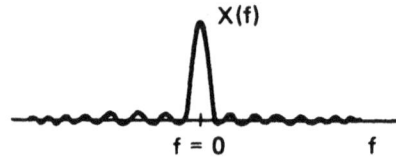

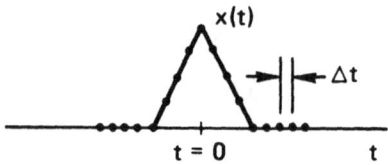

WE SAMPLE AT INCREMENTS OF Δt . . . .   AND OBTAIN A SPECTRUM
THAT IS REPETITIVE

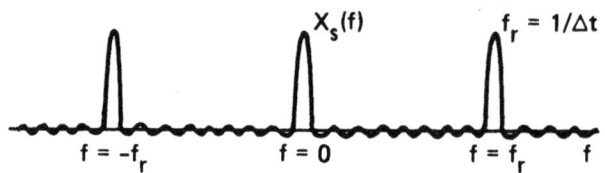

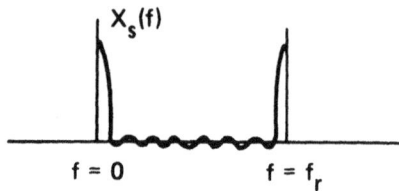

IF WE CREATE A NEW FUNCTION CONSISTING OF ONLY
ONE REPETITION INTERVAL OF $X_s(f)$ . . . .

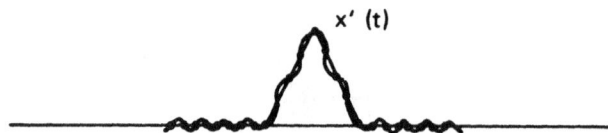

ITS FOURIER TRANSFORM IS COINCIDENT WITH $x(t)$
AT THE SAMPLES, BUT NOT IN BETWEEN

7.2/Sampling in the Frequency Domain

Because of the duality between signals in the time and frequency domains, there exists a corresponding duality between signals sampled in the time domain and signals sampled in the frequency domain. For example, we can approximate $x(t)$ by sampling the spectrum $X(f)$ at discrete frequencies $n\Delta f$ as

$$x_s(t) = \Delta f \sum_{n=-\infty}^{\infty} X(n\Delta f)\, e^{j2\pi t n \Delta f} \qquad (7.24)$$

The function $x_s(t)$ is periodic with a *repetition period* given by

$$t_r = 1/\Delta f \qquad (7.25)$$

It is related to the original function by

$$x_s(t) = \sum_{m=-\infty}^{\infty} x(t - mt_r) \qquad (7.26)$$

Since $x_s(t)$ is periodic, it will closely approximate $x(t)$ only if the repetition period t_r is larger than a time interval that contains most of the energy of the signal, which we designate as the duration T. Otherwise we get aliasing or foldover in the time domain, just as we got foldover in the frequency domain when we sampled signals in the time domain. Since

$$t_r \geqslant T \qquad (7.27)$$

then by the use of Eq. (7.25), we must have

$$\Delta f \leqslant 1/T \qquad (7.28)$$

Table 7.2/Fourier Transform Relationships for Signals Sampled in Time Domain

$$X_s(f) = \sum_{m=-\infty}^{\infty} X(f+mf_r)$$

$$X_s(f) = \Delta t \sum_{k=-\infty}^{\infty} x(k\Delta t)e^{-j2\pi f k \Delta t}$$

$$x(k\Delta t) = \int_{0}^{f_r} X_s(f)e^{j2\pi f k \Delta t}\, df \quad , f_r = 1/\Delta t$$

$$E_c = \Delta t \sum_{k=-\infty}^{\infty} |x(k\Delta t)|^2 = \int_{0}^{f_r} |X_s(f)|^2\, df$$

$$\Delta t \sum_{k=-\infty}^{\infty} e^{-j2\pi f k \Delta t} = \sum_{m=-\infty}^{\infty} \delta(f-mf_r)$$

$$\frac{1}{f_r} \int_{0}^{f_r} e^{j2\pi f k \Delta t}\, df = 1 \quad , k = 0$$
$$= 0 \quad , k \neq 0$$

if $\hat{x}(k\Delta t) = \Delta t \cdot x(k\Delta t)$, then

$$X_s(f) = \sum_{k=-\infty}^{\infty} \hat{x}(k\Delta t)e^{-j2\pi f k \Delta t}$$

$$\hat{x}(k\Delta t) = \frac{1}{f_r} \int_{0}^{f_r} X_s(f)e^{j2\pi f k \Delta t}\, df$$

$$\Delta t \cdot E_c = \sum_{k=-\infty}^{\infty} |\hat{x}(k\Delta t)|^2 = \frac{1}{f_r} \int_{0}^{f_r} |X_s(f)|^2\, df$$

The Fourier transform of $x_s(t)$ is the sequence of samples $X(n\Delta f)$. We can perform the Fourier transform only over one period of $x_s(t)$ to obtain a new function

$$X'(f) = \int_0^{t_r} x_s(t)\, e^{-j2\pi ft}\, dt \qquad (7.29)$$

If we change the subscript n in Eq. (7.24) to m and substitute for $x_s(t)$ in Eq. (7.29), we obtain

$$X'(f) = \sum_{m=-\infty}^{\infty} X(m\Delta f)\, \text{sinc}(m-ft_r)\, e^{j\pi(m-ft_r)} \qquad (7.30)$$

At the samples $f = n\Delta f$, this reduces to

$$X'(n\Delta f) = X(n\Delta f) = \int_0^{t_r} x_s(t)\, e^{-j2\pi tn\Delta f}\, dt \qquad (7.31)$$

which completes the derivation of the Fourier transform pair with Eq. (7.24).

The energy relationship corresponding to Eqs. (7.24) and (7.31) is given by

$$E_c = \int_0^{t_r} |x_s(t)|^2\, dt = \Delta f \sum_{n=-\infty}^{\infty} |X(n\Delta f)|^2 \qquad (7.32)$$

Since the Δf factor in Eqs. (7.24) and (7.32) is just a scale factor it is possible to combine it with the frequency sample as

$$\hat{X}(n\Delta f) = \Delta f \cdot X(n\Delta f) \qquad (7.33)$$

where the caret now designates a scaling of the frequency-domain function. Then Eq. (7.24) can be rewritten as

$$x_s(t) = \sum_{n=-\infty}^{\infty} \hat{X}(n\Delta f)\, e^{j2\pi tn\Delta f} \qquad (7.34)$$

and by making use of Eq. (7.25), we can write Eq. (7.31) as

$$\hat{X}(n\Delta f) = \frac{1}{t_r} \int_0^{t_r} x_s(t)\, e^{-j2\pi tn\Delta f}\, dt \qquad (7.35)$$

And finally, Eq. (7.32) can be rewritten as

$$P_{av} = \Delta f \cdot E_c = \frac{1}{t_r} \int_0^{t_r} |x_s(t)|^2\, dt = \sum_{n=-\infty}^{\infty} |\hat{X}(n\Delta f)|^2 \qquad (7.36)$$

Table 7.3/Fourier Transform Relationships for Signals Sampled in Frequency Domain

$$x_s(t) = \sum_{m=-\infty}^{\infty} x(t+mt_r)$$

$$X(n\Delta f) = \int_0^{t_r} x_s(t) e^{-j2\pi tn\Delta f}\, dt \quad, t_r = 1/\Delta f$$

$$x_s(t) = \Delta f \sum_{n=-\infty}^{\infty} X(n\Delta f) e^{j2\pi tn\Delta f}$$

$$E_c = \int_0^{t_r} |x_s(t)|^2\, dt = \Delta f \sum_{n=-\infty}^{\infty} |X(n\Delta f)|^2$$

$$\frac{1}{t_r} \int_0^{t_r} e^{-j2\pi tn\Delta f}\, dt = 1 \quad, n = 0$$
$$= 0 \quad, n \neq 0$$

$$\Delta f \sum_{n=-\infty}^{\infty} e^{j2\pi tn\Delta f} = \sum_{m=-\infty}^{\infty} \delta(t-mt_r)$$

if $\hat{X}(n\Delta f) = \Delta f \cdot X(n\Delta f)$, then

$$\hat{X}(n\Delta f) = \frac{1}{t_r} \int_0^{t_r} x_s(t) e^{-j2\pi tn\Delta f}\, dt$$

$$x_s(t) = \sum_{n=-\infty}^{\infty} \hat{X}(n\Delta f) e^{j2\pi tn\Delta f}$$

$$P_{av} = \Delta f \cdot E_c = \frac{1}{t_r} \int_0^{t_r} |x_s(t)|^2\, dt = \sum_{n=-\infty}^{\infty} |\hat{X}(n\Delta f)|^2$$

which we have designated as the average power within a repetition period. We summarize the results for sampling signals in the frequency domain in Table 7.3.

7.3/Sampling in Both Time and Frequency Domains

The next logical step to consider is what happens when signals are sampled in both the time and frequency domains as we must do for numerical solutions of the Fourier transform. First of all, let us begin with a continuous signal that is specified in the time domain as $x(t)$. We will sample this signal at the discrete times $k\Delta t$ and compute $X_s(f)$ by means of Eq. (7.3), but only at discrete frequencies. Since $X_s(f)$ is repetitive over an interval f_r, let us divide this interval into N_r equal increments with the first sample at $f = 0$ so that the frequency sampling increment is given by

$$\Delta f = f_r / N_r \qquad (7.37)$$

For the moment we will not specify how large N_r should be. Since $f_r = 1/\Delta t$ from Eq. (7.6), we can also write

$$\Delta f = 1/N_r \Delta t \qquad (7.38)$$

and

$$N_r = 1/\Delta f \Delta t \qquad (7.39)$$

In discrete notation Eq. (7.3) becomes

$$X_s(n\Delta f) = \Delta t \sum_k x(k\Delta t)\, e^{-j2\pi kn\Delta f\Delta t} \qquad (7.40)$$

where, for the moment, we will also not specify the range of the index k. In Eq. (7.40) the index n can be incremented over $n = 0, \ldots, N_r - 1$

without $X_s(n\Delta f)$ repeating. We can simplify Eq. (7.40) by the use of Eq. (7.39) as

$$X_s(n\Delta f) = \Delta t \sum_k x(k\Delta t)\, e^{-j2\pi kn/N_r} \qquad (7.41)$$

All we have done so far is to rewrite Eq. (7.3) to yield discrete samples of $X_s(f)$. But now that we have samples of $X_s(f)$, we note from Section 7.2 that a function sampled in the frequency domain is repetitive in the time domain. For a frequency sampling increment of Δf the repetition period is $t_r = 1/\Delta f$ as we have written in Eq. (7.25). Then from Eq. (7.38) we can also write

$$t_r = N_r \Delta t \qquad (7.42)$$

which means that the repetition period is also divided into N_r equal increments. So far we have not specified the interval over which $x(t)$ is sampled in Eq. (7.41). But if we are to make the Fourier transform of $X_s(n\Delta f)$ correspond to the samples of $x(t)$, then $x(t)$ must be time limited to a duration not greater than t_r. This condition means that there must not be more than N_r samples of the original signal in the summation in Eq. (7.41). Since we are apparently free to choose the size of N_r, it is this time-duration constraint that determines N_r, not some particular choice of the product $\Delta f \Delta t$ in Eq. (7.39). If $x(t)$ is not time-limited, then we must create a time-limited version of it that closely approximates $x(t)$ such that there is negligible energy lost in the limiting process. In the following we will assume that $x(t)$ is time-limited.

In Section 7.1 we derived the inverse Fourier transform for one repetition interval of the continuous function $X_s(f)$. For discrete samples of $X_s(f)$ let us rewrite Eq. (7.16) as a discrete summation as

$$x'(k\Delta t) = \Delta f \sum_{n=0}^{N_r-1} X_s(n\Delta f)\, e^{j2\pi kn/N_r} \qquad (7.43)$$

where we have used Eq. (7.39) to simplify the phase term. We note immediately that $x'(k\Delta t)$ is repetitive over N_r samples. We can now replace

Figure 7.3/Discrete Fourier Transform Pair Relationships

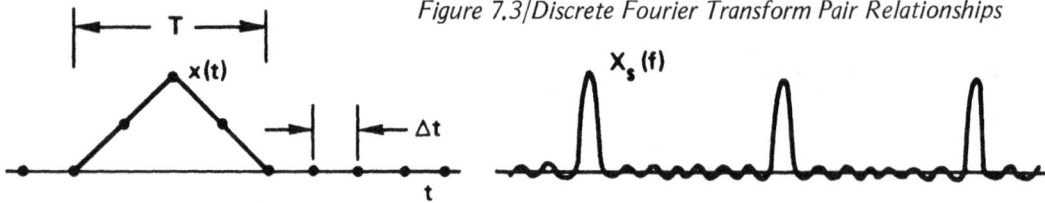

WE BEGIN WITH SAMPLES OF x(t) . . . .

AND OBTAIN A SPECTRUM THAT IS REPETITIVE

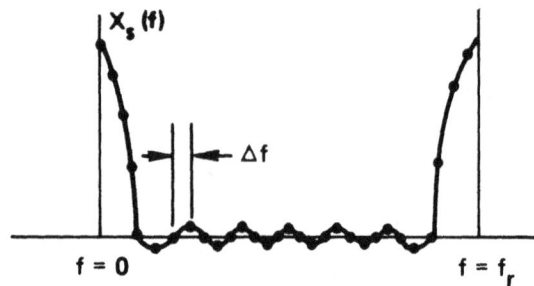

WE TAKE ONE REPETITION INTERVAL OF $X_s(f)$ AND SAMPLE IT AT INCREMENTS OF Δf . . . .

WHERE $N_r = f_r/\Delta f$ IS LARGER THAN $T/\Delta t$, THE NUMBER OF SAMPLES IN THE ORIGINAL FUNCTION x(t)

WE TAKE ITS FOURIER TRANSFORM AT DISCRETE INCREMENTS IN TIME . . . .

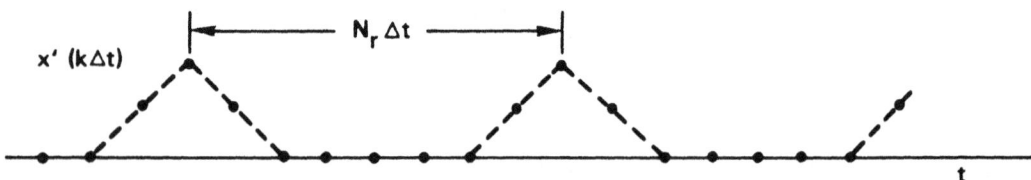

THE RESULT IS A REPETITIVE SEQUENCE OF SAMPLES THAT, OVER ONE REPETITION PERIOD, IS COINCIDENT WITH SAMPLES OF THE ORIGINAL FUNCTION.

$X_s(n\Delta f)$ in Eq. (7.43) by the right side of Eq. (7.41), but, as before, let us change the subscript k in Eq. (7.41) to m so that

$$x'(k\Delta t) = \Delta f \Delta t \sum_m x(m\Delta t) \sum_{n=0}^{N_r-1} e^{j2\pi n(k-m)/N_r} \qquad (7.44)$$

To aid in evaluating the summation over n we note that for an arbitrary u,

$$\sum_{n=0}^{N_r-1} u^n = \frac{1-u^{N_r}}{1-u} \qquad (7.45)$$

Now let

$$u = e^{j2\pi\ell/N_r} \qquad (7.46)$$

where ℓ is any integer. Hence

$$\sum_{n=0}^{N_r-1} e^{j2\pi\ell n/N_r} = \frac{1-e^{j2\pi\ell}}{1-e^{j2\pi\ell/N_r}} \qquad (7.47)$$

Since ℓ is an integer, the numerator is zero for all ℓ. The denominator is zero only for ℓ equal to zero or some integral multiple of N_r. At these values the left side is easily evaluated so that

$$\sum_{n=0}^{N_r-1} e^{j2\pi\ell n/N_r} = N_r, \quad \ell = 0, \pm N_r, \ldots$$
$$= 0, \qquad \text{otherwise} \qquad (7.48)$$

Since the summation over m in Eq. (7.44) is only over N_r consecutive samples, we can set $k-m = \ell$ in Eq. (7.48) and note that all terms in Eq. (7.44) vanish except for the m = k term. And since $N_r \Delta f \Delta t = 1$, Eq. (7.44) reduces to

$$x'(k\Delta t) = x(k\Delta t) \qquad (7.49)$$

where k assumes those values over which x(t) is defined, which we have already stated is not to exceed N_r samples. Thus Eq. (7.43) forms a discrete Fourier transform pair with Eq. (7.41), but with a repetitive function $x'(k\Delta t)$ that is coincident with samples of the original function $x(k\Delta t)$ for those values of k over which x(t) is defined. We show an example in Fig. 7.3 to illustrate these relationships.

So far we have allowed the sequence of samples of the original function x(t) to occur anywhere. But since the sequence $x'(k\Delta t)$ that belongs with the discrete Fourier transform pair is repetitive, it is convenient to assume that the original function is also repetitive over a period, $t_r = N_r\Delta t$, and choose the primary interval, $0 \leqslant t \leqslant t_r$, to be the one of interest. We illustrate how this is done in Fig. 7.4. Again for convenience we choose the samples at k = 0, ..., N_r-1, so that the *discrete Fourier transform pair* can be written as

$$X_s(n\Delta f) = \Delta t \sum_{k=0}^{N_r-1} x(k\Delta t) \, e^{-j2\pi kn/N_r} \qquad (7.50)$$
$$n = 0, \ldots, N_r-1$$

Figure 7.4/Reflection of Original Function to Primary Interval

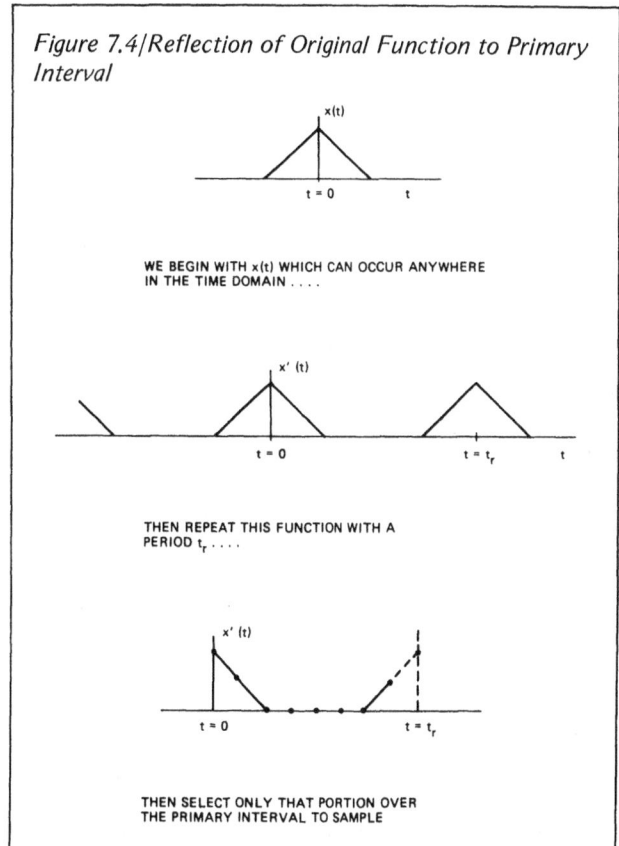

x(t)

t = 0   t

WE BEGIN WITH x(t) WHICH CAN OCCUR ANYWHERE IN THE TIME DOMAIN . . . .

x' (t)

t = 0   t = t_r   t

THEN REPEAT THIS FUNCTION WITH A PERIOD t_r . . . .

x' (t)

t = 0   t = t_r

THEN SELECT ONLY THAT PORTION OVER THE PRIMARY INTERVAL TO SAMPLE

$$x(k\Delta t) = \Delta f \sum_{n=0}^{N_r-1} X_s(n\Delta f)\, e^{j2\pi kn/N_r} \qquad (7.51)$$
$$k = 0, \ldots, N_r-1$$

We now make no distinction between $x(k\Delta t)$ and $x'(k\Delta t)$.

In this development we began with samples of the continuous, but time-limited, signal $x(t)$ and obtained samples of $X_s(f)$. We could also begin with samples of a continuous, but bandlimited, spectrum $X(f)$ and obtain samples of $x_s(t)$. Because of the duality between signals in the time and frequency domains we can easily derive the *discrete Fourier transform pair* for this case as

$$x_s(k\Delta t) = \Delta f \sum_{n=0}^{N_r-1} X(n\Delta f)\, e^{j2\pi kn/N_r} \qquad (7.52)$$
$$k = 0, \ldots, N_r-1$$

$$X(n\Delta f) = \Delta t \sum_{k=0}^{N_r-1} x_s(k\Delta t)\, e^{-j2\pi kn/N_r} \qquad (7.53)$$
$$n = 0, \ldots, N_r-1$$

The important point to remember is that we must begin by specifying the function in only one domain since the result of the discrete Fourier transform is a function that only approximates the Fourier transform of the original function.

7.4/Simplified Notation

The discrete Fourier transform pair in Eqs. (7.50) and (7.51) is identical in form to the pair in Eqs. (7.52) and (7.53), except that the subscript s is switched from the frequency-domain function $X_s(n\Delta f)$ to the time-domain function $x_s(k\Delta t)$. To simplify the notation let us drop the subscript, although not the implicit meaning, to obtain a single *discrete Fourier transform pair* as

$$X(n\Delta f) = \Delta t \sum_{k=0}^{N_r-1} x(k\Delta t)\, e^{-j2\pi kn/N_r} \qquad (7.54)$$
$$n = 0, \ldots, N_r-1$$

$$x(k\Delta t) = \Delta f \sum_{n=0}^{N_r-1} X(n\Delta t)\, e^{j2\pi kn/N_r} \qquad (7.55)$$
$$k = 0, \ldots, N_r-1$$

The appearance of Δt and Δf outside the summations in Eqs. (7.54) and (7.55) is somewhat unhandy in discrete computations since they are really just scale factors. These factors are necessary, however, to make the equations dimensionally correct. If $x(t)$ has the dimension of volts, then $X(f)$ has the dimension volt/Hz or volt-sec. An alternate definition of the sampling process in time is to combine the Δt factor with $x(t)$ as we did in Eq. (7.20), which we rewrite in abbreviated form as

$$\hat{x}_k = \Delta t \cdot x(k\Delta t) \qquad (7.56)$$

The definition results in the more convenient discrete Fourier transform pair which can be derived from Eqs. (7.54) and (7.55) as

$$X_n = \sum_{k=0}^{N_r-1} \hat{x}_k\, e^{-j2\pi kn/N_r}, \qquad n = 0, \ldots, N_r-1 \qquad (7.57)$$

$$\hat{x}_k = \frac{1}{N_r} \sum_{n=0}^{N_r-1} X_n\, e^{j2\pi kn/N_r}, \quad k = 0, \ldots, N_r-1 \qquad (7.58)$$

where $X_n = X(n\Delta f)$ and we have used Eq. (7.39) to obtain the $1/N_r$ factor in Eq. (7.58). Now X_n and $\hat{x}_k$ have the same dimensions. Similarly, we can combine the Δf factor with $X(f)$ as in Eq. (7.33), which we rewrite in abbreviated form as

$$\hat{X}_n = \Delta f \cdot X(n\Delta f) \qquad (7.59)$$

This quantity has the interpretation of voltage within a frequency band of width Δf. This definition results in a slightly different, but equivalent form for the discrete Fourier transform pair as

$$\hat{X}_n = \frac{1}{N_r} \sum_{k=0}^{N_r-1} x_k\, e^{-j2\pi kn/N_r}, \qquad n = 0, \ldots, N_r-1 \quad (7.60)$$

$$x_k = \sum_{n=0}^{N_r-1} \hat{X}_n\, e^{j2\pi kn/N_r}, \qquad k = 0, \ldots, N_r-1 \quad (7.61)$$

where $x_k = x(k\Delta t)$; we have again used Eq. (7.39) to obtain the $1/N_r$ factor in Eq. (7.60). Note that in the transform pair of Eqs. (7.57) and (7.58) the $1/N_r$ factor appears in the reverse direction, while in Eqs. (7.60) and (7.61) it appears in the forward direction. In the remaining discussion, we will use the caret ˆ to denote the scaling of the frequency-domain variable by Δf and the time-domain variable by Δt. We summarize these results in Table 7.4.

For the energy-type relationship for signals sampled in both domains we write the *average power* within a repetition period as

$$P_{av} = \frac{1}{N_r} \sum_{k=0}^{N_r-1} |x_k|^2 \qquad (7.62)$$

To express the average power in terms of samples of the spectrum we substitute Eq. (7.61) into Eq. (7.62) and rearrange the order of summation to obtain

$$P_{av} = \frac{1}{N_r} \sum_{m=0}^{N_r-1} \sum_{n=0}^{N_r-1} \hat{X}_m \hat{X}_n^* \sum_{k=0}^{N_r-1} e^{j2\pi(m-n)/N_r} \qquad (7.63)$$

By the use of Eq. (7.48) the summation over k is non-zero only for m = n, so Eq. (7.63) reduces to

$$P_{av} = \sum_{n=0}^{N_r-1} |\hat{X}_n|^2 \qquad (7.64)$$

In this expression and in Eq. (7.62) both $|x_k|^2$ and $|\hat{X}_n|^2$ have the same dimensions of power. We can interpret $|\hat{X}_n|^2$ as the spectral power within a frequency interval Δf. Similarly, by using Eqs. (7.57) and (7.58) we can derive

---

Table 7.4/Discrete Fourier Transform Relationships

$$x_k = x(k\Delta t) \qquad \hat{x}_k = \Delta t \cdot x(k\Delta t)$$

$$X_n = X(n\Delta f) \qquad \hat{X}_n = \Delta f \cdot X(n\Delta f)$$

$$X_n = \sum_{k=0}^{N_r-1} \hat{x}_k\, e^{-j2\pi kn/N_r}$$

$$\hat{x}_k = \frac{1}{N_r} \sum_{n=0}^{N_r-1} X_n\, e^{j2\pi kn/N_r} \qquad f_r = 1/\Delta t$$

$$t_r = 1/\Delta f$$

$$\sum_{k=0}^{N_r-1} |\hat{x}_k|^2 = \frac{1}{N_r} \sum_{n=0}^{N_r-1} |X_n|^2 \qquad N_r = f_r/\Delta f$$

$$= t_r/\Delta t$$

$$\hat{X}_n = \frac{1}{N_r} \sum_{k=0}^{N_r-1} x_k\, e^{-j2\pi kn/N_r} \qquad = 1/\Delta f \Delta t$$

$$= f_r\, t_r$$

$$x_k = \sum_{n=0}^{N_r-1} \hat{X}_n\, e^{j2\pi kn/N_r}$$

$$P_{av} = \frac{1}{N_r} \sum_{k=0}^{N_r-1} |x_k|^2 = \sum_{n=0}^{N_r-1} |\hat{X}_n|^2$$

$$\sum_{n=0}^{N_r-1} e^{j2\pi \ell n/N_r} = N_r \quad , \quad \ell = 0, \pm N_r, \pm 2N_r, \ldots$$

$$= 0 \quad , \quad \text{otherwise}$$

$$\sum_{k=0}^{N_r-1} |\hat{x}_k|^2 = \frac{1}{N_r} \sum_{n=0}^{N_r-1} |X_n|^2 \qquad (7.65)$$

which has no direct relationship to average power. Although $|\hat{x}_k|^2$ and $|X_n|^2$ have the same dimensions, only $|X_n|^2$ is conveniently interpreted as samples of energy spectral density.

7.5/Brute-Force and Fast Fourier Transforms

The discrete Fourier transform pair given by Eqs. (7.57) and (7.58), or by Eqs. (7.60) and (7.61), is in a form that permits the highly efficient Cooley-Tukey algorithm to be used. The result is the so-called *Fast Fourier Transform* (FFT) where a sequence of length N_r is transformed into another sequence of length N_r. All N_r points must be specified on input, even though many might be zero, and we obtain all N_r points on output, even though we might need fewer. The computation time is generally proportional to N_r times the sum of the prime factors of N_r. Note that we do not have to constrain N_r to be 2 raised to an integer power, although such a number would ensure a fast transformation. Only in this case is the computation time proportional to $2N_r \log_2 N_r$. Otherwise, we should make N_r rich in the prime factors of 2, 3, and 5.

For many applications it is faster to implement the discrete Fourier transform directly, especially if many of the input samples are zero and we need only a relatively few samples on output. We will designate this implementation as the *brute-force transform* since we are not taking advantage of any special algorithms to speed up the computation. The computation time is proportional to the product of the number of samples on input and the number of samples on output. If both of these numbers are N_r, then the computation time is proportional to N_r^2, which is clearly less efficient than the Fast Fourier Transform.

To illustrate the difference between the fast and brute-force transform we will compare assembly-language subroutines written for the CDC6400 computer. For the brute-force transform, N_f and N_t are the number of complex samples used in the frequency and time domains, respectively. The execution times are

FFT – (20 μsec) $N_r \log_2 N_r$

Brute Force – (40 μsec) $N_f N_t$

where the FFT time is based on an algorithm ($N_r = 2^{integer}$) by Singleton [Ref. 6]. As an example, suppose that $N_f = 5$, $N_t = 12$, and $N_r = 64$. The FFT would require 3.2 msec while the brute-force transform would require only 2.4 msec.

7.6/Filtering and Convolution

In Section 4.4 we defined the operation of filtering and the resultant Fourier transform domain operation of convolution in terms of continuous functions. Since it does not matter which function we identify with the waveform or filter, let us simply denote the operation of filtering in terms of arbitrary functions as

$$Z(f) = X(f)Y(f) \qquad (7.66)$$

where the corresponding Fourier transform pairs $\{X(f), x(t)\}$, $\{Y(f), y(t)\}$, and $\{Z(f), z(t)\}$ exist. The time domain convolution is given by

$$z(t) = \int_{-\infty}^{\infty} x(t')y(t-t')dt' \tag{7.67}$$

If we sample $x(t)$ and $y(t)$ at discrete intervals spaced by Δt, we can approximate $z(t)$ by

$$z_s(k\Delta t) = \Delta t \sum_m x(m\Delta t)y[(k-m)\Delta t] \tag{7.68}$$

where the summation is over all samples for which the functions $x(t)$ and $y(t)$ overlap. Here, we have developed the discrete convolution as an approximation to the continuous function. If we use the shorthand notation for the convolution, we can also write Eq. (7.68) as

$$z_k = x_k \star y_k \tag{7.69}$$

where it is understood that $\{z_k\}$ refers to the sequence $\{z_s(k\Delta t)\}$. We will designate Eq. (7.68) or Eq. (7.69) as the direct method of implementing a discrete convolution, in contrast to another method we will discuss.

Let us define the following discrete functions which have discrete Fourier transform pairs — $\{\hat{x}_k, X_n\}$, $\{\hat{y}_k, Y_n\}$, and $\{\hat{z}_k, Z_n\}$. We are using the same notation as in Eqs. (7.57) and (7.58) where the caret ^ denotes a scaling of the time variable by Δt as in Eq. (7.56). Now let

$$Z_n = X_n Y_n \tag{7.70}$$

If we substitute the discrete Fourier transforms of X_n and Y_n into Eq. (7.70), we obtain

$$Z_n = \sum_{m=0}^{N_r-1} \sum_{\ell=0}^{N_r-1} \hat{x}_m \hat{y}_\ell \, e^{-j2\pi(m+\ell)n/N_r} \tag{7.71}$$

The discrete Fourier transform of Eq. (7.70), after we rearrange the order of summation, is given by

$$\hat{z}_k = \frac{1}{N_r} \sum_{m=0}^{N_r-1} \sum_{\ell=0}^{N_r-1} \hat{x}_m \hat{y}_\ell \sum_{n=0}^{N_r-1} e^{-j2\pi(m+\ell-k)n/N_r}$$

By the use of Eq. (7.48) the summation over n is non-zero for $\ell = k - m$ so that we can now write

$$\hat{z}_k = \sum_{m=0}^{N_r-1} \hat{x}_m \hat{y}_{k-m} , \quad k = 0, \ldots, N_r-1 \tag{7.72}$$

It appears that we could have derived this result directly by multiplying both sides of Eq. (7.68) by Δt. However, there is one fundamental difference between Eqs. (7.72) and (7.68). The functions in Eqs. (7.72) are periodic, whereas those in Eq. (7.68) are not. Note that when $k = 0$ in Eq. (7.72), the subscripts on $\hat{y}$ are negative. This poses no problem since $\hat{y}_{k-N_r} = \hat{y}_k$, but it is more convenient to write the equivalent expression

$$\hat{z}_k = \sum_{m=0}^{N_r-1} \hat{x}_m \hat{y}_{(k-m)(\text{mod } N_r)} , \quad k = 0, \ldots, N_r-1 \tag{7.73}$$

where all functions have subscripts only in the interval $(0, N_r-1)$. This result can be interpreted as a *discrete circular convolution*. We can also interchange the subscripts on $\hat{x}$ and $\hat{y}$ as

$$\hat{z}_k = \sum_{m=0}^{N_r-1} \hat{y}_m \hat{x}_{(k-m)(\text{mod } N_r)}, \quad k = 0, \ldots, N_r-1 \tag{7.74}$$

In any case the operations in Eqs. (7.72) through (7.74) are equivalent. In the shorthand notation for the convolution, we will distinguish the circular nature of the convolution as ⊛ so that Eqs. (7.73) and (7.74) can be written as

$$\hat{z}_k = \hat{x}_k ⊛ \hat{y}_k \tag{7.75}$$

There are alternate ways to write the discrete filtering and convolution relationships that involve scaling in the frequency domain by Δf. For example, if we multiply both sides of Eq. (7.70) by Δf, we can lump the Δf on the right side with either X_n or Y_n to form $\hat{X}_n$ or $\hat{Y}_n$, respectively. The corresponding time-domain relationships are found by dividing both sides of Eq. (7.72) by Δt.

Table 7.5/Comparison of Continuous and Discrete Convolutions

	Continuous	Discrete
Frequency-domain filtering	$Z(f) = X(f)Y(f)$	$Z_n = X_n Y_n$
Time-domain filtering	$z(t) = x(t) \star y(t)$	$\hat{z}_k = \hat{x}_k \circledast \hat{y}_k$

Discrete convolution-direct method

$$z_s(k\Delta t) = \Delta t \sum_m x(m\Delta t)y[(k-m)\Delta t]$$

	Continuous	Discrete
Time-domain weighting	$z(t) = x(t)y(t)$	$z_k = x_k y_k$
Frequency-domain convolution	$Z(f) = X(f) \star Y(f)$	$\hat{Z}_n = \hat{X}_n \circledast \hat{Y}_n$

We can also obtain convolution relationships in the frequency domain if we multiply two time-domain functions. The relationships are the duals of those obtained for convolution in the time domain, with time and frequency being interchanged, as we show in Table 7.5.

7.7/Implementing the Convolution

There are two approaches to implementing the discrete convolution. The *direct method* is by means of Eq. (7.68). The *indirect method*, given the time-domain samples $x(k\Delta t)$ and $y(k\Delta t)$, is as follows:

1/ Compute the sequences $\{X_n\}$ and $\{Y_n\}$ by the discrete Fourier transformation of $\{\hat{x}_k\}$ and $\{\hat{y}_n\}$.

2/ Form $Z_n = X_n Y_n$ for $n = 0, \ldots, N_r - 1$.

3/ Compute the sequence $\{\hat{z}_k\}$ by the discrete Fourier transformation of $\{Z_n\}$.

For the indirect method all discrete Fourier transformations must be performed over N_r samples, assuming we use the FFT.

Although the indirect method seems to involve more effort, it is actually more efficient to implement in certain situations. Suppose the number of complex samples in $x(t)$ is N_x, the number in $y(t)$ is N_y, and we want N_z samples of $z(t)$ where $N_z = N_x + N_y$. The computation time for the direct method is proportional to the product $N_x N_y$. On the CDC6400 computer, the execution time for complex samples is $(40\ \mu\text{sec})\ N_x N_y$.

Eqs. (7.70) and (7.74) are the discrete operations that correspond to the time-domain filtering and convolution in Eqs. (7.66) and (7.67), respectively, as we summarize in Table 7.5. Note that Eq. (7.68) has no equivalent discrete operation in the frequency domain. The difference between Eqs. (7.68) and (7.74) is illustrated in Fig. 7.5, where we note that all functions involved in Eq. (7.74) are periodic, whereas those in Eq. (7.68) are not. If one condition is met, however, the two operations will be equivalent. This condition is that N_r, the number of samples in the repetition interval, must be at least as large as the total number of samples in $\{z_k\}$; otherwise repeated versions of the sequence $\{z_k\}$ will overlap. If N_x and N_y are the number of samples in $\{x_k\}$ and $\{y_k\}$, respectively, to prevent overlap we must fill out the remainder of each array with zeros to a total length of N_r samples where

$$N_r \geqslant N_x + N_y \qquad (7.76)$$

In terms of the continuous functions we must satisfy

$$t_r \geqslant T_x + T_y \qquad (7.77)$$

where T_x and T_y are the durations of $x(t)$ and $y(t)$, respectively.

Figure 7.5/Comparison of Direct and Discrete Circular Convolutions

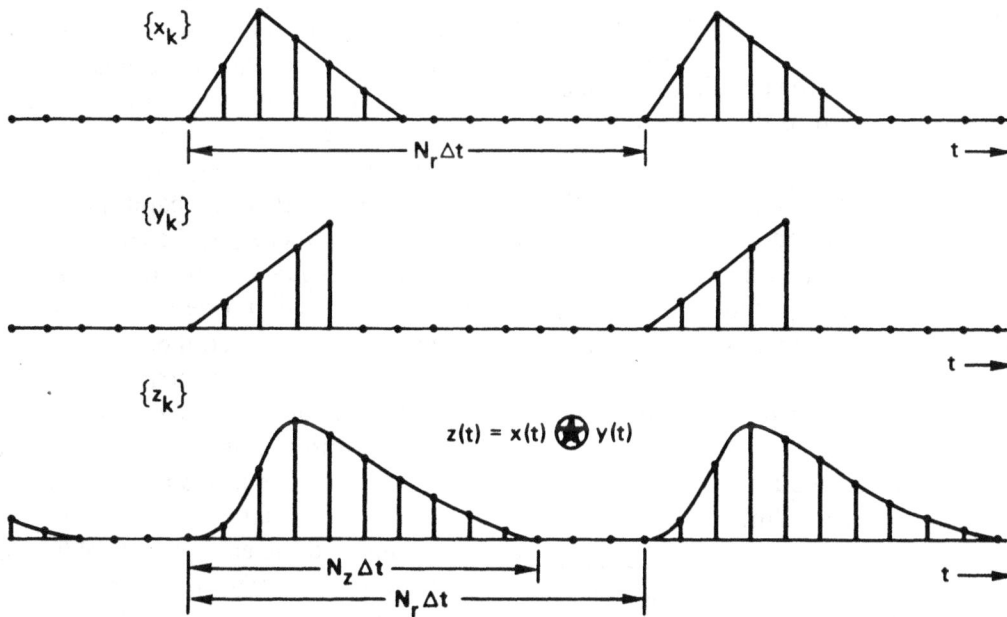

DIRECT DISCRETE CONVOLUTION:

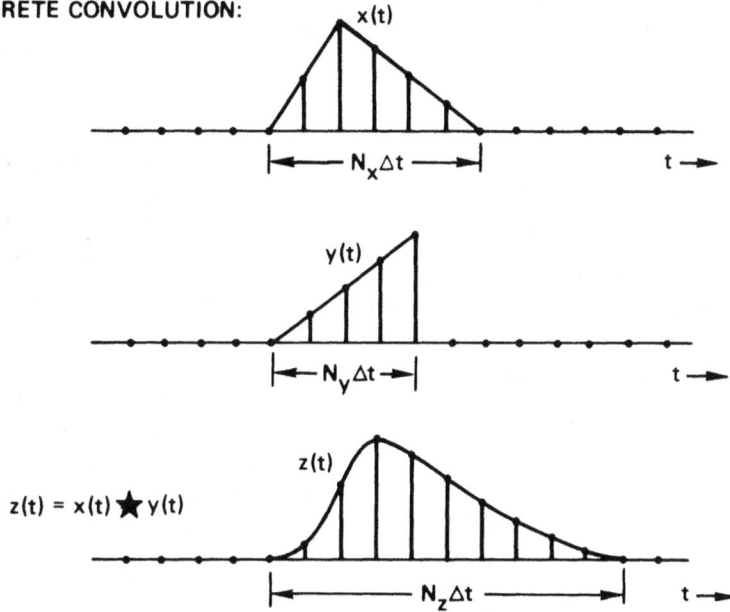

x(t)

$N_x \Delta t$

y(t)

$N_y \Delta t$

z(t) = x(t) ★ y(t)

z(t)

$N_z \Delta t$

DISCRETE CIRCULAR CONVOLUTION:

$\{x_k\}$

$N_r \Delta t$

$\{y_k\}$

$\{z_k\}$

z(t) = x(t) ✪ y(t)

$N_z \Delta t$

$N_r \Delta t$

If we use the Fast Fourier Transform for the indirect method, we must select $N_r \geq N_z$. In general there are three transforms involved* so that the execution time on the CDC6400 computer is (60 μsec) $N_r \log_2 N_r$ with $N_r = 2^{integer}$. As an example, let $N_x = N_y = 1000$ and $N_r = 2048$. The direct method will require 40 sec, compared to the indirect method which requires only 1.35 sec. As both N_x and N_y get smaller, or if either one is small, it may be more advantageous to use the direct method. If $N_x = 10$, $N_y = 100$, and $N_r = 128$, the direct method requires 40 msec compared to the indirect method which requires 54 msec. Note that if $N_x = N_y$ and $N_r = 2N_x$, the breakeven point is reached when $N_x = 32$. This condition is true not only for the CDC6400, but also for the GE630 [Ref.1].

The comparison becomes more favorable for the direct method if not all possible samples of z(t) are desired, as is often the case in correlation analysis. If N_z is the number of samples desired, the execution time on the CDC6400 is (40 μsec) N_z min (N_x, N_y) for complex samples.

The computation time for the indirect method remains unchanged. For the previous example of $N_x = N_y = 1000$ and $N_r = 2048$, suppose N_z is now only 10. The time for the direct method is only 0.40 sec compared to the 1.35 sec for the indirect method.

In many cases one of the functions x(t) or y(t) will be real. This fact also reduces the number of computations for the direct method, but it does not substantially change the indirect method. If all $N_z = N_x + N_y$ samples are desired, the execution time on the CDC6400 for the discrete convolution of a complex sequence with a real one is (20 μsec) $N_x N_y$.

There are other fast techniques that can be used to implement discrete convolutions as discussed

---

*Sometimes either x(t) or y(t) will already be transformed so that the indirect method would require only two FFT's.

by Stockham [Ref. 5 (pp. 330-334)]. These techniques generally apply when one function is considerably longer than the other, especially when the shortest length function is still reasonably long.

7.8/Sampling Rates

We have already specified inequalities that must be satisfied when signals are sampled in either the time or frequency domain. In general the choice of sampling rates is straightforward for bandlimited signals sampled in the time domain (or for time-limited signals sampled in the frequency domain). Since a bandlimited signal will be of infinite duration (and consequently will not be physically realizable), we must be sure to include enough samples in the time domain so that there will be negligible energy in those portions of the time-domain signal not sampled.

Actually the principal case of interest is one where a time-limited signal is sampled in the *time* domain (or the dual case of a bandlimited signal sampled in the frequency domain). Here we get repetition and foldover in the spectrum where none existed for the original continuous signal. As we indicated earlier, the repetition frequency must be larger than the band that contains most of the spectral energy; the larger we make the repetition frequency, the lower the energy in the foldover region. Energy there is clearly related to how the spectral energy is distributed. We can show this by referring to Fig. 7.1b. Within each repetition frequency interval the energy that folds into that interval from the spectral component on the left is the portion of the spectral energy that is above $f_r/2$ in the original spectrum, and the energy that folds into that interval from the right is that portion below $-f/2$ in the original spectrum. Or equivalently, the total energy that folds into each repetition interval is just the energy that lies outside of the band of width f_r centered on the mean frequency of the original spectrum. Obviously the larger we make f_r, the less energy there will be outside of this band and the less energy that folds into each repetition frequency interval.

The distribution of spectral energy can be used as a quantitative measure of the error due to sampling. If the energy that folds into each repetition frequency interval is, say 10^{-4} of the total signal energy, then the relative errors associated with sampling will also be of the order of 10^{-4}. The actual measure of sampling error depends on how the samples will be used. But to avoid treating each case as a special case we will simply say that the sampling error is measured by the relative energy that folds into each repetition frequency interval, assuming that we are interested in the whole interval. We will give an example later.

To evaluate this error due to sampling it is convenient to compute the *spectral distribution of energy*. First we define the distribution of spectral energy within a band of width F centered on the mean (or reference) frequency f_o as

$$d(F) = \int_{-F/2}^{F/2} |X(f-f_o)|^2 \, df \qquad (7.78)$$

Then we define the more useful quantity

$$D(F) = 1 - d(F)/d(\infty) \qquad (7.79)$$

which is the relative spectral energy outside of the band of width F. Thus the sampling error is measured by $D(f_r)$, assuming that we are interested in the whole repetition frequency interval.

As an example, let us refer to Fig. 7.1a where the power spectrum is a Gaussian-shaped function as

$$|X(f)|^2 = e^{-(\alpha f/f_{3dB})^2} \qquad (7.80)$$

where f_{3dB} is the two-sided half-power width and

$$\alpha = 2\sqrt{\ln 2} = 1.665 \qquad (7.81)$$

is a constant that satisfies $|X(f_{3dB}/2)|^2 = 0.5$. The relative spectral energy outside a band of width F can be written in terms of the complementary error function as

$$D(F) = \text{erfc}(\alpha F/2f_{3dB}) \qquad (7.82)$$

We show a plot of this function in Fig. 7.6.

In some cases we will not be interested in the whole repetition frequency interval. For example, we might low-pass filter the sampled data so that most of the energy in the foldover region is suppressed. Or we might be interested in only a portion of the spectrum. Thus it is expedient to define this new band of interest as being of width F_o (which incidentally can be centered on any arbitrary frequency; f_o in Eq. (7.78) will designate this reference frequency). Now let us redraw Fig. 7.1b as in Fig. 7.7 where we have indicated this band. Each edge of this band is at a distance $f_r - F_o/2$ from the reference frequency of the adjacent spectrum. Since all the repeated spectra are identical, the energy entering the spectral band of interest from the two adjacent repeated spectra is identical to the energy outside the band of the original spectrum of width $2(f_r - F_o/2) = 2f_r - F_o$ centered on the reference frequency. The *relative* energy that folds into the band of interest is then $D(2f_r - F_o)$. There are also contributions from repeated spectra further away, but these are completely negligible. If the width of the spectral band of interest is the repetition frequency ($F_o = f_r$), then the relative energy that folds into this band is $D(f_r)$, as we have already shown. For the Gaussian-shaped spectrum used to derive the curve in Fig. 7.6, we can easily see that if $F = 2f_r - F_o \geq 4f_{3dB}$, the relative energy that folds into the band of width F_o will be almost 10^{-6}. If $F_o = f_r$, then this inequality reduces to $f_r \geq 4f_{3dB}$.

Figure 7.6/Spectral Energy Distribution for Gaussian-
Shaped Spectrum

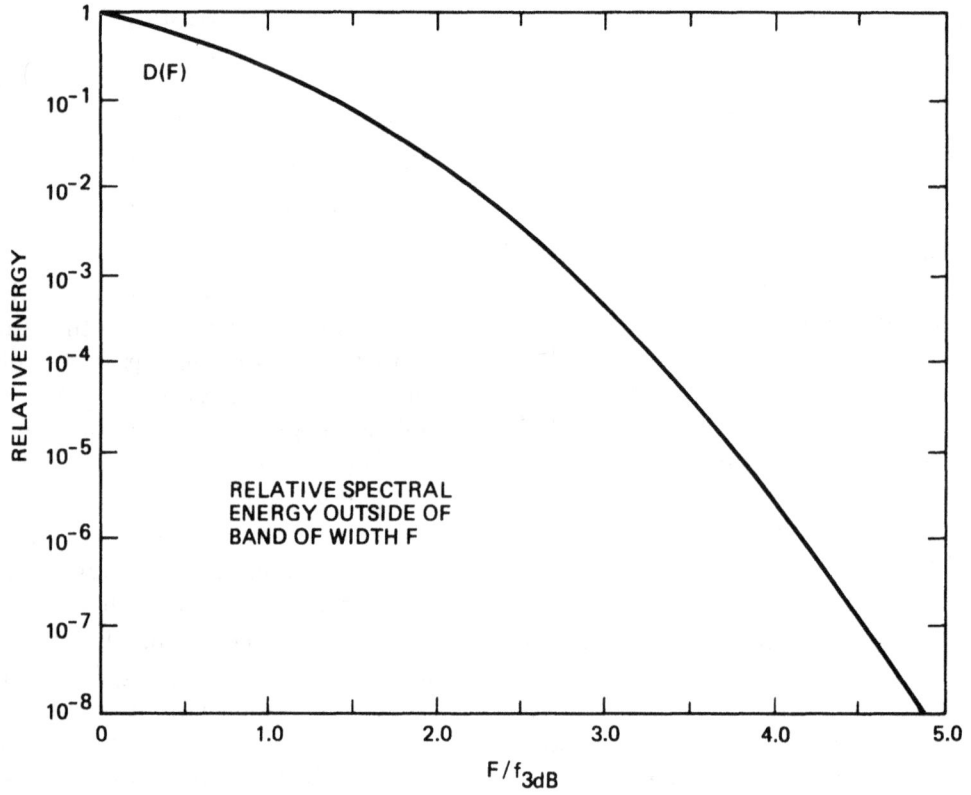

D(F)

RELATIVE ENERGY

10^{-1}
10^{-2}
10^{-3}
10^{-4}
10^{-5}
10^{-6}
10^{-7}
10^{-8}

RELATIVE SPECTRAL
ENERGY OUTSIDE OF
BAND OF WIDTH F

0   1.0   2.0   3.0   4.0   5.0

F/f_{3dB}

Figure 7.7/Spectrum of the Sampled Signal

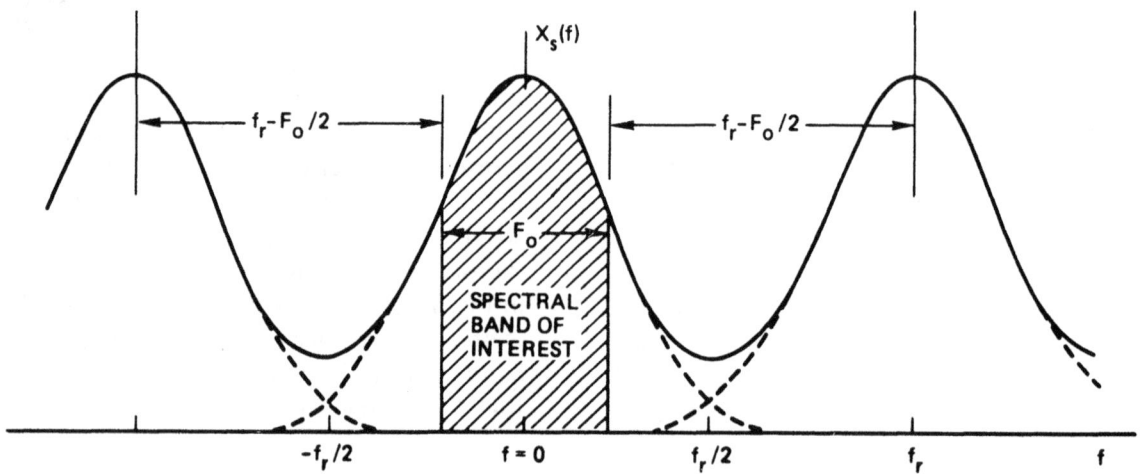

$X_s(f)$

$f_r - F_o/2$

$f_r - F_o/2$

F_o

SPECTRAL
BAND OF
INTEREST

$-f_r/2$ $f = 0$ $f_r/2$ f_r f

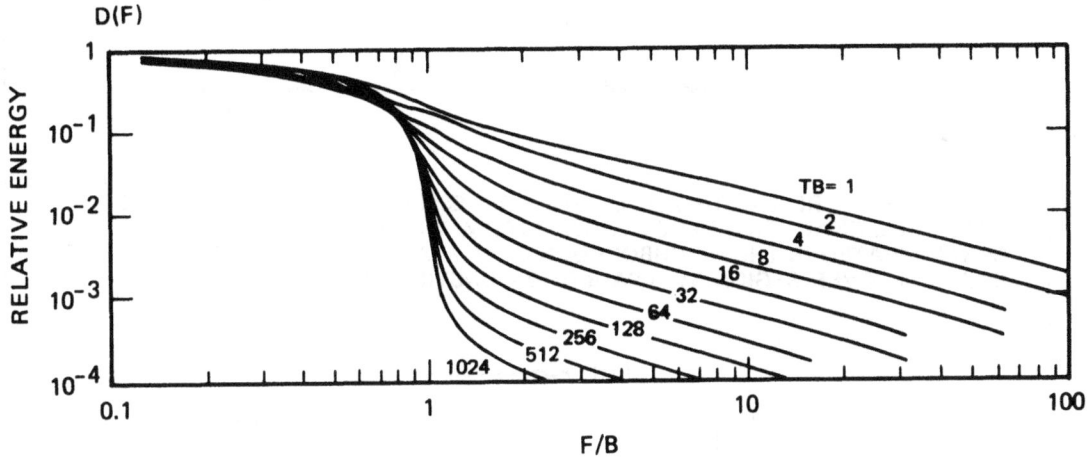

Figure 7.8/Spectral Energy Distribution for Linear-FM
Signal with Uniformly Weighted Envelope (relative energy outside of band of width F where B is signal bandwidth)

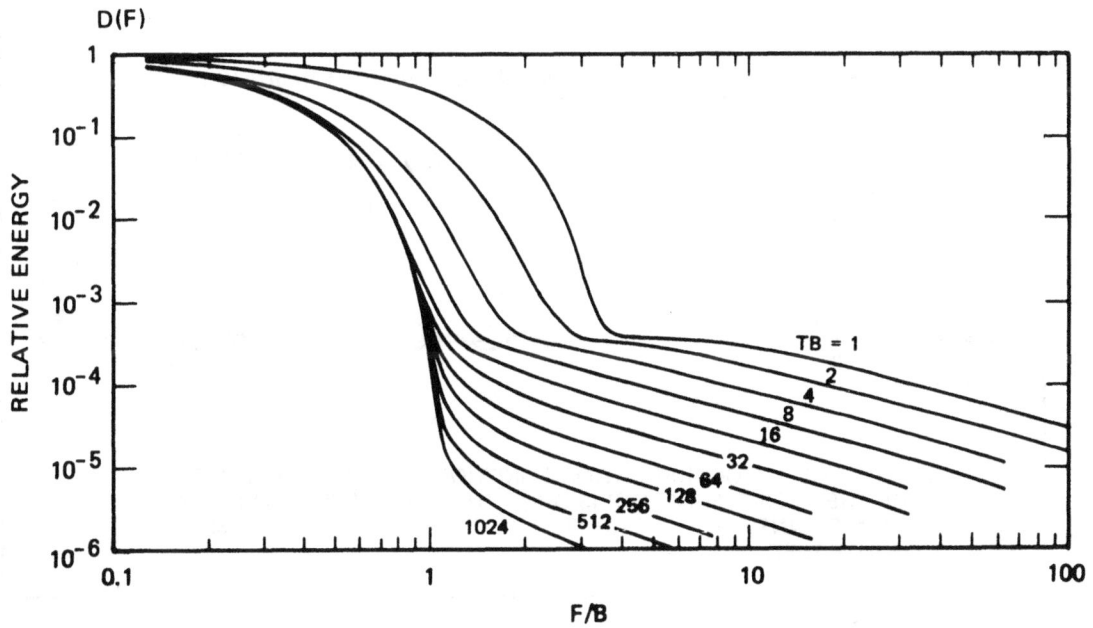

Figure 7.9/Spectral Energy Distribution for Linear-FM
Signal with Hamming Weighted Envelope

Figure 7.10/Effect of Sampling Error on Resulting Autocorrelation Function

WE BEGIN WITH THE ENERGY DENSITY SPECTRUM OF SOME ARBITRARY SIGNAL CONSTRAINED TO HAVE UNIT ENERGY

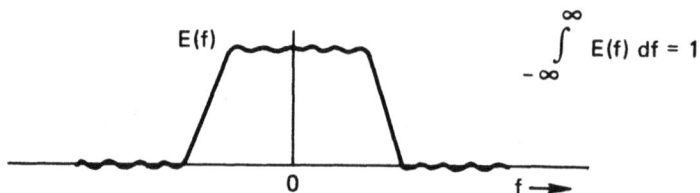

$$\int_{-\infty}^{\infty} E(f)\, df = 1$$

IF WE WERE TO REPEAT THIS SPECTRUM, CORRESPONDING TO SAMPLING THE ORIGINAL SIGNAL, THE ENERGY ASSOCIATED WITH FOLDOVER WILL BE CONCENTRATED IN THE NEIGHBORHOOD OF $-f_r/2$ AND $f_r/2$. LET US APPROXIMATE EACH FOLDOVER REGION BY A PAIR OF δ-FUNCTIONS AS

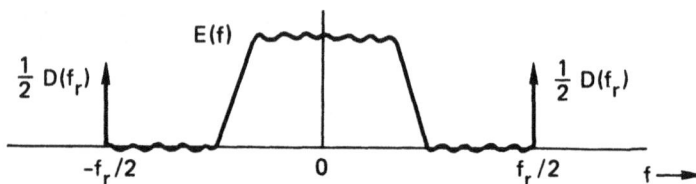

THE FOURIER TRANSFORM OF THIS COMPOSITE SPECTRUM RESULTS IN TWO ADDITIVE FUNCTIONS AS

1. THE AUTOCORRELATION FUNCTION

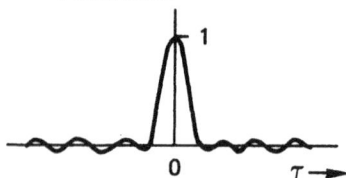

2. A COSINE FUNCTION

THE PEAK OF THE AUTOCORRELATION FUNCTION IS UNITY AND THE PEAK HEIGHT OF THE COSINE FUNCTION IS $D(f_r)$. THUS THE COSINE FUNCTION WILL BE SIGNIFICANT ONLY IF $D(f_r)$ IS LARGER THAN THE LOWEST SIDELOBE LEVEL OF INTEREST.

The spectral energy distribution is strongly dependent on both the shape of the signal envelope in the time domain and the time-bandwidth product of a phase modulated signal, which we illustrate by selecting the linear-FM signal (see Section 5.4) as an example. In Fig. 7.8 we show the distribution of spectral energy $D(F)$ for the linear-FM signal where the time-bandwidth product TB is a parameter. We observe that the spectral energy is much more concentrated about the mean frequency for large time-bandwidth products than for small ones. For example, a relatively small repetition frequency of $f_r = 2B$ for TB = 128 results in about the same relative energy in the foldover region ($F = f_r$) as does the large repetition frequency of $f_r = 100B$ for TB = 2. In Fig. 7.9 we have repeated the conditions used for Fig. 7.8 except that the envelope on the linear-FM signal is Hamming weighted. We observe the same general effect of the spectral energy being more concentrated about the mean frequency as the time-bandwidth product is increased. However, for this case the relative energies in the sidelobe region are about two orders of magnitude lower.

In Section 6.4 we have already discussed some of the effects of sampling. In general we said that we could isolate mainbeam effects (targets) from sidelobe effects (clutter) in the process of mapping scatterers, since the two effects were essentially independent. This distinction allows us to utilize rather coarse sampling intervals when mapping scatterers; an error of a few percent will have a negligble effect on the simulation results. But now we are sampling signals. We will see that an error of a few percent in the sampling process may not be tolerable since the signal sidelobes may be affected considerably. Let us consider the example of computing the autocorrelation function of a sampled signal. As we illustrate in Fig. 7.1, sampling a continuous signal will cause some energy to foldover at the frequencies $f = \pm f_r/2$. Let us combine all of this foldover energy into δ-functions at $f = \pm f_r/2$, as we show in Fig. 7.10. The resulting autocorrela-

tion function will be the autocorrelation function of the original continuous signal plus a cosine function. The peak height of the cosine function relative to the peak of the autocorrelation function is $D(f_r)$. If the level of this cosine function is lower than the original sidelobe levels, then there will be a negligible effect on the discrete computation of the autocorrelation function. To illustrate this point with a specific example, we have computed the autocorrelation function of a sampled linear-FM signal of time-bandwidth product TB = 32 and $f_r/B = 8$ and compared this with the continuous autocorrelation function given by Eq. (5.47) for $\nu = 0$. The difference is plotted in Fig. 7.11, from which we note a peak error of about 3×10^{-4}. From Fig. 7.8 we have $D(F = f_r) = 10^{-3}$, which is somewhat larger than the actual error. This result is to be expected since the δ-function used above is only an approximation to the foldover energy that actually is distributed over some spectral interval.

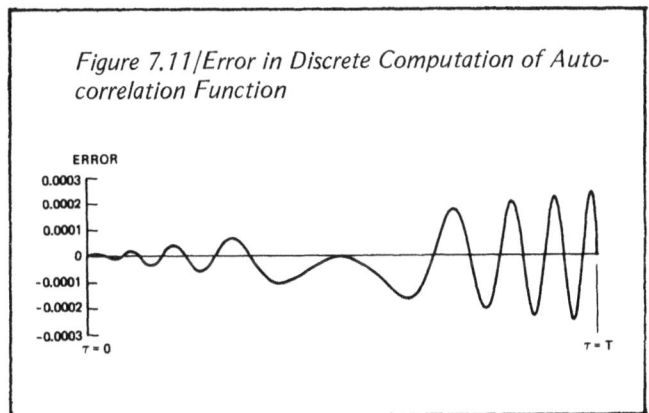

Figure 7.11/Error in Discrete Computation of Autocorrelation Function

A general rule of thumb that can be used in sampling signals is to choose a sampling rate so that sampling error is about equal to the lowest sidelobe level of interest (relative to the mainlobe peak). Lower sampling rates will cause too much distortion in the sidelobe levels while higher sampling rates will be inefficient to implement in large simulations. For the linear-FM signal with a rectangular envelope, this fact might dictate a sampling error of about 10^{-2} since the sidelobes of the autocorrelation function are rather high. From Fig. 7.8 we see that an acceptable sampling rate for TB = 32 that would yield $D(f_r)$ = 10^{-2} would be about f_r/B = 1.5. For the Hamming weighted envelope, the sampling error should be about 10^{-4} since the sidelobes of the autocorrelation function are comparable to this value. From Fig. 7.9 we see that an acceptable sampling rate for TB = 32 would again be about f_r/B = 1.5. The fact that these two sampling rates are the same is not coincidental. The two sets of curves in Figs. 7.8 and 7.9 would be virtually identical if we translated the set in Fig. 7.8 downward by 18 dB. Since most spectral sidelobes of the Hamming weighted signal are about 42 dB down from the mainlobe peak, we will use this level as a reference. The corresponding level for the rectangular envelope signal is -24 dB. A sampling rate based on these two levels would be independent of the signal envelope. In Table 7.6 we list the minimum (and optimum) sampling rates for the linear-FM signal as a function of the time-bandwidth product TB, assuming we are interested in the whole spectral interval of width f_r (i.e., we solve for

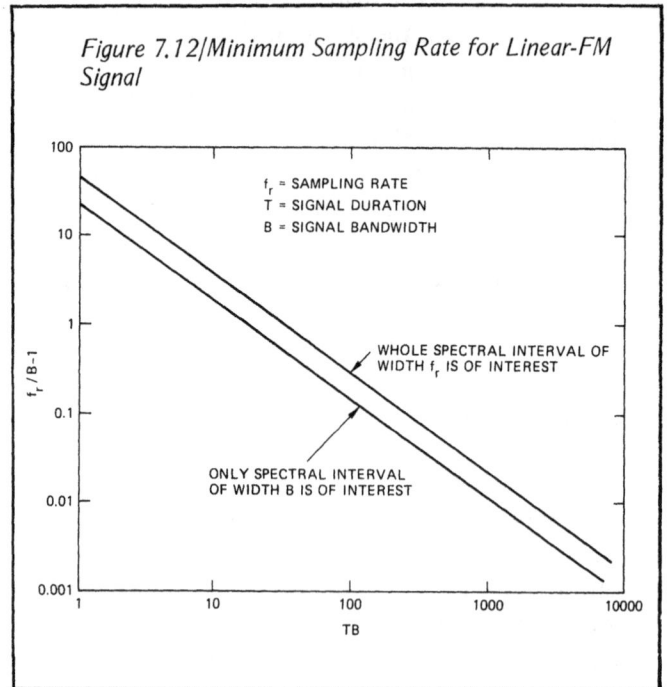

Figure 7.12/Minimum Sampling Rate for Linear-FM Signal

$D(f_r)$ = -42 dB in the case of the Hamming weighted signal). The use of these values will ensure that the sampling errors will have a negligible effect on the signal responses. In Fig. 7.12 we show $f_r/B - 1$ plotted as a straight line as a function of TB on suitably chosen scales. Values derived from Table 7.6 or Fig. 7.12 are probably applicable for all radar signals of interest and not just the linear-FM signal.

In many cases it will not be necessary to keep the sampling errors down over the whole spectral interval of width f_r. For example, we may only be interested in the spectral interval corresponding to the signal bandwidth. In this case we can solve for $D(2f_r - B)$ = -42 dB for the Hamming weighted signal. Given the value of f_r/B in Table 7.6, we can easily compute the new value of f_r/B as 0.5 plus 0.5 times the value in the table. We list these new values in Table 7.7 and also plot them as the lower curve in Fig. 7.12. Note that these sampling rates are significantly lower than those in Table 7.6 for the smaller time-bandwidth products.

Table 7.6/Minimum Sampling Rates for Linear-FM Signal (whole spectral interval of width f_r is of interest)

TB	f_r/B	f_rT
1	46	46
2	21	42
4	10.3	41
8	5.9	47
16	3.2	51
32	1.90	61
64	1.47	94
128	1.22	156
256	1.10	282
512	1.045	535
1024	1.019	1043
2048	1.005	2058

Table 7.7/Minimum Sampling Rates for Linear-FM Signal (only spectral interval corresponding to signal bandwidth is of interest)

TB	f_r/B	f_rT
1	24	24
2	11	22
4	5.7	23
8	3.5	28
16	2.1	34
32	1.45	47
64	1.24	79
128	1.11	142
256	1.05	269
512	1.023	524
1024	1.010	1034
2048	1.003	2053

7.9/Sampling Real Signals

So far we have discussed only complex signals. Real signals are characterized by a spectral envelope that is symmetrical about f = 0, as was shown in Fig. 4.1. The above results can also be applied to real signals if we designate the mean frequency as f = 0. Or we can derive the one-sided distribution of spectral energy as

$$d_1(F) = \int_0^F |X(f)|^2 \, df \qquad (7.83)$$

and the relative energy above f = F as

$$D_1(F) = 1 - d_1(F)/d_1(\infty) \qquad (7.84)$$

In Fig. 7.13 we show another example of the repeated spectrum of a sampled real signal. Let us designate the highest frequency of interest as F_1, again assuming that we are not interested in

what happens for frequencies higher than F_1. We see that this frequency is at a distance of $f_r - F_1$ from the first repetition frequency. The energy in the left side of the repeated spectrum that falls into the frequency band of interest will be determined by $D_1(f_r - F_1)$. The largest useable band of interest is obviously $F_1 = f_r/2$, which leads us to the Nyquist sampling rate for real signals, $f_r \geqslant 2F_1$, and a sampling error that is determined by $D_1(f_r/2)$. Usually F_1 is designated as the maximum frequency of the spectrum such that there is negligible energy beyond F_1, and then f_r is chosen so that $f_r \geqslant 2F_1$.

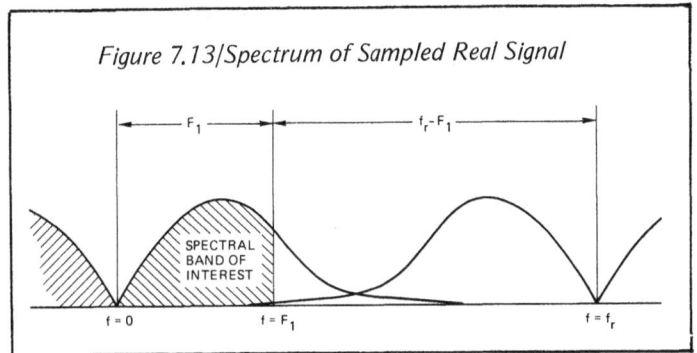

Figure 7.13/Spectrum of Sampled Real Signal

Let us compare the sampling rates for a real signal and for a complex signal derived from the same real one by the methods of Section 4.2. The signal duration in either case will be designated as T and the highest frequency of interest as F_1. For the real signal, the error due to sampling will be measured by $D_1(f_r - F_1)$. If we allow for a different sampling rate, f_r', for the complex signal, the error will be measured by $D(2f_r' - F_1)$ as we derived in Section 7.8, assuming that we are interested in the whole band from f = 0 to f = F_1. Now we can equate Eqs. (7.83) and (7.78) by defining the reference frequency in Eq. (7.78) to be $f_o = F_1/2$. Thus we see that $2f_r' = f_r$ or $f_r' = f_r/2$. In other words the sampling rate for the complex signal will be half of the sampling rate for the real signal, with the resulting errors due to sampling being identical. Note that the number of samples in the real signal is Tf_r, and the number of complex samples in the complex signal is $Tf_r' = Tf_r/2$. But a complex sample has two components, so the total number of scalar samples for either case is Tf_r.

In some cases the minimum frequency of a real signal will not be f = 0. As far as the real signal is concerned, nothing has changed in determining the sampling rate. The error due to sampling will still depend on F_1, the highest frequency of interest. However, the situation is different for the complex signal. The frequency band of interest will be less than F_1, which means that the sampling rate for the complex signal can be less than $f_r/2$ and still achieve the same sampling error. We will need fewer total samples if we work with the complex signal.

References

[1]   Brigham, E.O.; *The Fast Fourier Transform*; Prentice-Hall, Englewood Cliffs, New Jersey, 1974.

[2]   Bergland, G.D.; "A Guided Tour of the Fast Fourier Transform," *IEEE Spectrum, Vol. 6, No. 7*, July 1969, pp. 41-52.

[3]   *IEEE Trans. Audio and Electroacoustics — Special Issue on Fast Fourier Transform and Its Application to Digital Filtering and Spectral Analysis — Vol. AU-15*, June 1967.

[4]   *IEEE Trans. Audio and Electroacoustics — Special Issue on Fast Fourier Transform — Vol. AU-17*, June 1969.

[5]   Rabiner, L.R. and C.M. Rader (eds.); *Digital Signal Processing*; IEEE Press, New York, 1972.

[6]   Singleton, R.C.; "On Computing the Fast Fourier Transform," *Communications ACM, Vol. 10, No. 10*, October 1967, pp. 647-654. (The execution times quoted refer to Singleton's assembly-language FFT subroutine for the CDC6400/6600.)

Chapter 8

Generating Random Sequences

Most radar simulations (in fact, all Monte Carlo simulations) require the generation of random sequences. In this chapter we will give several methods of generating random sequences on a digital computer, where either the power spectral density or the autocorrelation function can be arbitrarily specified. We will confine the nature of the random process to that common in radar.

8.1/The Sampled Process

For a sampled or discrete random process we could apply the results of Chapter 7 directly to the Fourier transform pair in Eqs. (4.67) and (4.68) which relates the autocorrelation function and power spectral density. In contrast to the continuous random process, it is straightforward to derive the properties of a discrete random process directly.

Let us begin with a sequence of N_r complex random noise samples $\{x_k\}$ in the time domain. Although we will restrict our interest to these N_r samples, it is convenient to assume that the sequence repeats outside this interval every N_r samples. This repeated sequence now qualifies as a sampled signal in the sense that a discrete Fourier transform pair exists in the form derived in Chapter 7. If we use the convention in Eqs. (7.60) and (7.61), we can write

$$\hat{X}_n = \frac{1}{N_r} \sum_{k=0}^{N_r-1} x_k\, e^{-j2\pi kn/N_r}, \qquad n=0,\ldots,N_r-1 \quad (8.1)$$

$$x_k = \sum_{n=0}^{N_r-1} \hat{X}_n\, e^{j2\pi kn/N_r}, \qquad k=0,\ldots,N_r-1 \quad (8.2)$$

where $\{\hat{X}_n\}$ is defined as the spectral noise samples. This sequence is also repetitive every N_r samples. We choose to use this convention of combining the frequency sampling

increment with the spectral sample since both x_k and $\hat{X}_n$ will have the same dimensions of voltage.

We can develop the second-moment properties by first taking the absolute-value squared of Eq. (8.1) as

$$|\hat{X}_n|^2 = \frac{1}{N_r^2} \sum_{k=0}^{N_r-1} \sum_{\ell=0}^{N_r-1} x_k x_\ell^*\, e^{-j2\pi(k-\ell)n/N_r} \quad (8.3)$$

Since $\{x_k\}$ is repetitive every N_r samples, we can perform the summation in ℓ over any N_r consecutive samples. In particular we can rewrite the limits of the summations as

$$\sum_{k=0}^{N_r-1} \sum_{\ell=0}^{N_r-1} = \sum_{k=0}^{N_r-1} \sum_{\ell=k-N_r+1}^{k} \quad (8.4)$$

Furthermore, we can set $m = k-\ell$ in Eq. (8.3) to obtain

$$|\hat{X}_n|^2 = \frac{1}{N_r^2} \sum_{k=0}^{N_r-1} \sum_{m=0}^{N_r-1} x_k x_{k-m}^*\, e^{-j2\pi mn/N_r} \quad (8.5)$$

Now let us take the ensemble average of Eq. (8.5) as

$$\overline{|\hat{X}_n|^2} = \frac{1}{N_r^2} \sum_{k=0}^{N_r-1} \sum_{m=0}^{N_r-1} \overline{x_k x_{k-m}^*}\, e^{-j2\pi mn/N_r} \quad (8.6)$$

The ensemble average quantity within the summation is recognized as the autocorrelation function of the sequence $\{x_k\}$.

Let us abbreviate the notation of Eq. (4.58) to define the autocorrelation function of a discrete sequence as

$$R_m = \overline{x_k x_{k-m}^*} \quad (8.7)$$

If we substitute Eq. (8.7) into Eq. (8.6) the quantities within the summations do not involve the subscript k. Hence, we can write

$$\overline{|\hat{X}_n|^2} = \frac{1}{N_r} \sum_{m=0}^{N_r-1} R_m\, e^{-j2\pi mn/N_r} \quad (8.8)$$

Since the random noise sequence $\{x_k\}$ has the dimension of voltage, the autocorrelation function has the dimension of power as we can see from Eq. (8.7). Formally, we shall define the *power spectral sequence* $\{\hat{S}_n\}$ as

$$\hat{S}_n = \overline{|\hat{X}_n|^2} \tag{8.9}$$

where $\hat{S}_n$ represents the ensemble averaged power within the increment Δf.

Both R_m and S_n are repetitive every N_r samples — the former, by view of $\{x_k\}$ in Eq. (8.7) originally being constrained to be repetitive; the latter, by the fact that $\{X_n\}$ in Eqs. (8.8) and (8.9) is also repetitive. If we use the procedures in Chapter 7 we can define the discrete Fourier transform pair

$$\hat{S}_n = \frac{1}{N_r} \sum_{m=0}^{N_r-1} R_m \, e^{-j2\pi mn/N_r} \tag{8.10}$$

$$R_m = \sum_{n=0}^{N_r-1} \hat{S}_n \, e^{j2\pi mn/N_r} \tag{8.11}$$

$$m=0, \ldots, N_r-1$$

The ensemble averaged power is given by

$$P = R_o = \sum_{n=0}^{N_r-1} \hat{S}_n \tag{8.12}$$

In developing these discrete Fourier transform results we have avoided the use of limits that were required for the continuous functions. These discrete quantities are related to the continuous quantities in the same way that the discrete signals in Chapter 7 are related to the continuous signals. To relate $\hat{S}_n$ to the continuous quantity we note that

$$\hat{S}(n\Delta f) = \Delta f \cdot S(n\Delta f) \tag{8.13}$$

However, we must remember that the sampled functions in Eqs. (8.10) and (8.11) cannot be simultaneously equal to samples of the continuous functions in Eqs. (4.67) and (4.68).

8.2/Spectral Independence of Samples

One property of all discrete random processes is that spectral samples are independent, a fact which greatly simplifies the generation of random sequences. To show this independence, let us write the product of spectral samples in Eq. (8.1) as

$$\hat{X}_m \hat{X}_n^* = \frac{1}{N_r^2} \sum_{i=0}^{N_r-1} \sum_{\ell=0}^{N_r-1} x_i x_\ell^* \, e^{-j2\pi(im-\ell n)/N_r} \tag{8.14}$$

Now let us take the ensemble average as

$$\overline{\hat{X}_m \hat{X}_n^*} = \frac{1}{N_r^2} \sum_{i=0}^{N_r-1} \sum_{\ell=0}^{N_r-1} \overline{x_i x_\ell^*} \, e^{-j2\pi(im-\ell n)/N_r} \tag{8.15}$$

Just as we rewrote the summation in Eq. (8.4), we will write Eq. (8.15) with $k = i-\ell$ as

$$\overline{\hat{X}_m \hat{X}_n^*} = \frac{1}{N_r^2} \sum_{i=0}^{N_r-1} \sum_{k=0}^{N_r-1} \overline{x_i x_{i-k}^*} \, e^{-j2\pi(im-in+kn)/N_r} \tag{8.16}$$

We can rearrange the double summation and make use of Eq. (8.7) as

$$\overline{\hat{X}_m \hat{X}_n^*} = \frac{1}{N_r} \sum_{i=0}^{N_r-1} e^{-j2\pi i(m-n)/N_r} \cdot \frac{1}{N_r} \sum_{k=0}^{N_r-1} R_k \, e^{-j2\pi kn/N_r} \tag{8.17}$$

The first summation is non-zero only for $m = n$ by virtue of Eq. (7.48). Hence,

$$\overline{\hat{X}_m \hat{X}_n^*} = \frac{1}{N_r} \sum_{k=0}^{N_r-1} R_k \, e^{-j2\pi kn/N_r} , \qquad m=n$$

$$= 0 \qquad\qquad\qquad , \text{ otherwise} \tag{8.18}$$

which is also in agreement with Eq. (8.8).

In the next three sections we will show how to generate random sequences.

8.3/General Methods of Generating Random Sequences

We have just demonstrated that the spectral samples of a discrete random process are independent. Thus if we are given the power spectral sequence $\{\hat{S}_n\}$, we can generate one member of the ensemble of that random process by generating an independent random sequence $\{\hat{X}_n\}$ such that the ensemble average power is

$$\overline{|\hat{X}_n|^2} = \hat{S}_n \qquad (8.19)$$

as in Eq. (8.9). Note that the phase is arbitrary, and hence random over 2π radians. Thus the quadrature components must have a mean value of zero.

In Chapter 9 we will show how to generate a random phasor with a particular amplitude distribution. Let $\{\xi_n\}$ be an independent sequence of such phasors, where the ensemble average power is unity as

$$\overline{|\xi_n|^2} = 1 \qquad (8.20)$$

Now we can set

$$\hat{X}_n = \xi_n \sqrt{\hat{S}_n} \quad , n=0, 1, \ldots, N_r-1 \qquad (8.21)$$

so that the ensemble average power will still be given by Eq. (8.19).

After we have generated the independent random sequence in the frequency domain given by Eq. (8.21), we can create the corresponding correlated time sequence by a discrete Fourier transform as

$$x_k = \sum_{n=0}^{N_r-1} \hat{X}_n \, e^{j2\pi kn/N_r} \quad , k=0, 1, \ldots, N_r-1 \qquad (8.22)$$

The autocorrelation function of $\{x_k\}$ will be related to the power spectral sequence $\{\hat{S}_n\}$ by the discrete Fourier transform pair in Eqs. (8.10) and (8.11). We summarize the above steps in Fig. 8.1. With this approach there is no restriction on the probability distribution function for amplitude (the second moment must exist, however).

We have not specified how to choose N_r or Δf, though. Since we are dealing with a process that repeats in both the time and frequency domains, we know that the autocorrelation function of $\{x_k\}$ must be repetitive every N_r samples as we sketch in Fig. 8.2. Suppose the correlation time (defined loosely as the time for the process to become decorrelated) is given by $N_c \Delta t$. The maximum lag that we can have before we encounter correlation from the next repeated component is $(N_r-N_c)\Delta t$. Thus we must choose N_r large enough so that, for a maximum length sequence of N, $(N_r-N_c)\Delta t \geqslant N\Delta t$ or

$$N_r \geqslant N + N_c \qquad (8.23)$$

In order to avoid foldover of the repeated components in Fig. 8.2 we must have $N_r \geqslant 2 N_c$. Thus another lower bound on N_r can be written as

$$N_r \geqslant 2 \max \{N, N_c\} \qquad (8.24)$$

In some cases it is more convenient to use Eq. (8.24), even though it is not as close a bound as Eq. (8.23). To determine the frequency sample spacing we note that $\Delta f = 1/N_r\Delta t$.

Figure 8.1/Steps to Generate a Correlated Time Sequence.

BEGIN WITH THE POWER SPECTRAL SEQUENCE $\{\hat{S}_n\}$ AS

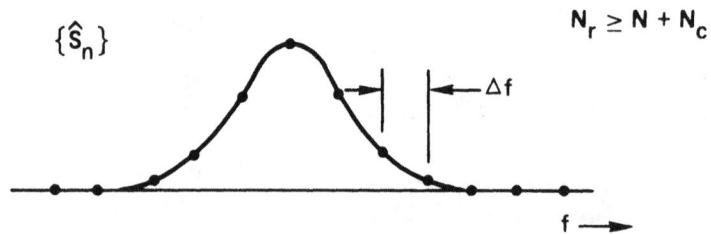

$\{\hat{S}_n\}$

$N_r \geq N + N_c$

Δf

$f \longrightarrow$

GENERATE AN INDEPENDENT SEQUENCE OF RANDOM PHASORS $\{\xi_n\}$ AS

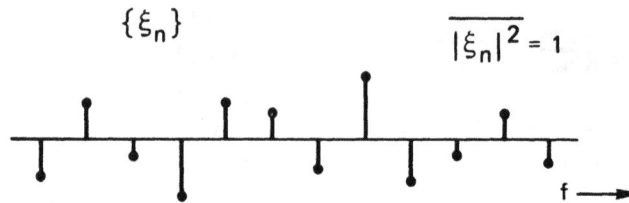

$\{\xi_n\}$

$\overline{|\xi_n|^2} = 1$

$f \longrightarrow$

(REAL PART SHOWN)

THEN SCALE EACH RANDOM PHASOR BY THE AMPLITUDE AS

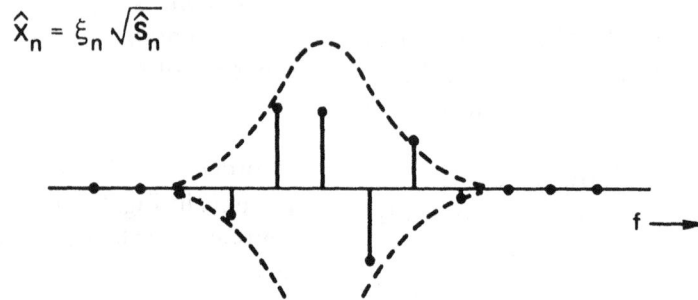

$\hat{x}_n = \xi_n \sqrt{\hat{S}_n}$

$f \longrightarrow$

TO CREATE THE TIME SEQUENCE TAKE THE DISCRETE FOURIER TRANSFORM AS

$$x_k = \sum_{n=0}^{N_r-1} \hat{x}_n \, e^{j2\pi kn/N_r} \quad , \; k = 0, \ldots , N_r-1$$

We can derive an alternate approach to generating random sequences. To do so let us treat the multiplication in Eq. (8.21) as a discrete filter by writing

$$\hat{H}_n = \sqrt{\hat{S}_n} \qquad (8.25)$$

We lose no generality by restricting $\{\hat{H}_n\}$ to be a real sequence. Now we can write

$$\hat{X}_n = \xi_n \hat{H}_n \qquad (8.26)$$

From Section 7.6 we see that this product in the frequency domain corresponds to a discrete convolution in the time domain as

$$x_k = \hat{\zeta}_k \star h_k \qquad (8.27)$$

where $\hat{\zeta}_k$ and h_k are the discrete Fourier transforms of ξ_n and $\hat{H}_n$, respectively. Thus Eq. (8.27) provides us with an alternate method of generating a random sequence, as we outline in Fig. 8.3. But this method is attractive only as long as we do not have to use a discrete Fourier transform to obtain $\{\hat{\zeta}_k\}$ from $\{\xi_n\}$; otherwise it would be as efficient to implement the product in Eq. (8.21) and the discrete Fourier transform in Eq. (8.22).

For the Gaussian process (here we refer to the statistics of each phasor) we do not have to explicity transform $\{\xi_n\}$ to $\{\hat{\zeta}_k\}$. If the quadrature components (the real and imaginary parts) of $\{\xi_n\}$ are Gaussian, the quadrature components of $\{\hat{\zeta}_k\}$ will be as well. Furthermore, because of Eq. (8.20), the time-domain sequence $\{\hat{\zeta}_k\}$ will be an independent random sequence. Thus for the Gaussian process we can generate the sequence $\{\hat{\zeta}_k\}$ directly in the time domain. To generate the correlated time sequence $\{x_k\}$ we just implement the discrete convolution in Eq. (8.27).

Which brings up the question as to how the random process is specified if it is non-Gaussian. If we specify the amplitude distribution of the frequency-domain sequence $\{\hat{X}_n\}$, then it is straightforward to generate a correlated time-domain sequence with the use of Eq. (8.22), regardless of the amplitude distribution function. However, if we specify the amplitude distribution of the time-domain sequence $\{x_k\}$, then it is not always straightforward to generate such a sequence. A correlated log-normal sequence is tractable to generate because it is a simple transformation of a correlated Gaussian sequence

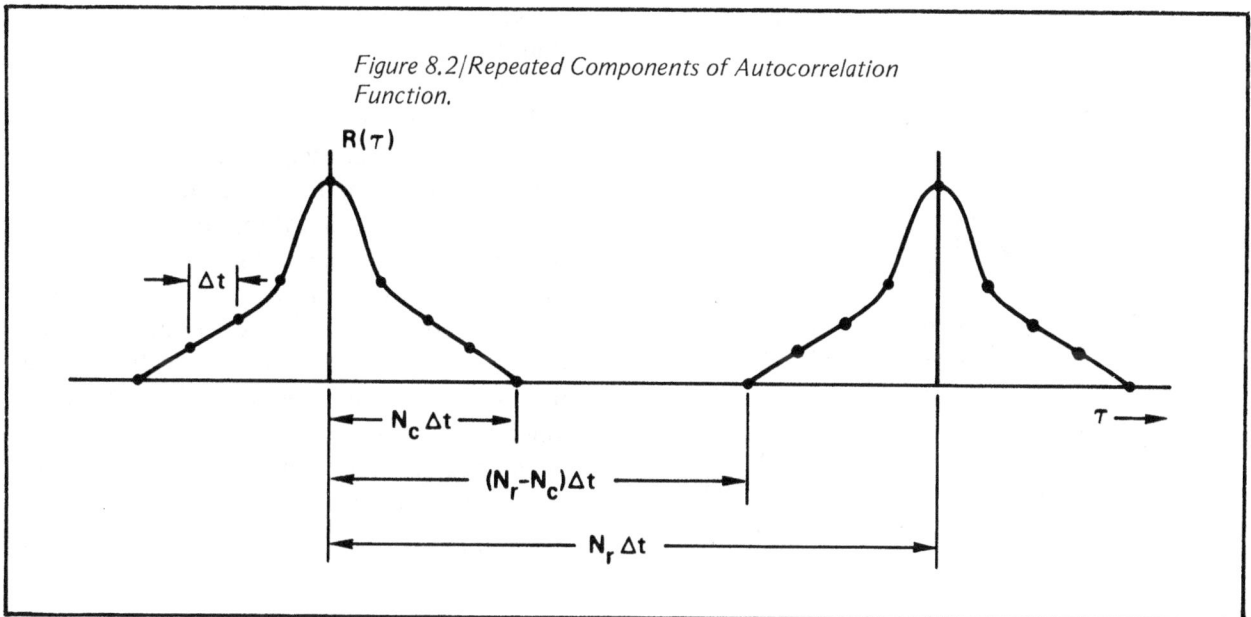

Figure 8.2/Repeated Components of Autocorrelation Function.

where the correlation properties of one sequence are easily written in terms of the other. Broste [Ref. 1] describes a general method of transforming a correlated Gaussian sequence into any desired non-Gaussian sequence, but an iterative procedure is required to obtain the correlation properties of the input sequence in terms of those of the desired output sequence. His method is an approximation and no procedure is easily derived for estimating the residual spectral errors. This consideration is important for simulating high-performance radars and, because of the difficulties involved in using Broste's or any similar method, we should always attempt to relate the physical process to the spectral, not the time-domain, quantities. Fortunately most radar models are expressed in spectral terms and are compatible with the methods discussed in this chapter, even for non-Gaussian processes.

8.4/Fast Algorithms Based on Gaussian-Shaped Spectrum

Many spectral models used in radar simulations will be Gaussian-shaped, or nearly so; we can gain much insight into the various methods of generating random sequences if we begin with such a spectrum. Let us define the Gaussian-shaped spectrum (the power spectral density) in terms of the two-sided half-power width f_{3dB} as

$$S(f) = e^{-(\alpha f/f_{3dB})^2} \qquad (8.28)$$

where α is a constant chosen so that $S(f_{3dB}/2) = 0.5$. Thus

$$\alpha = 2\sqrt{\ln 2} = 1.665 \qquad (8.29)$$

The autocorrelation function corresponding to $S(f)$ is the Fourier transform of Eq. (8.28) as

$$R(\tau) = e^{-(\pi f_{3dB}\tau/\alpha)^2} \qquad (8.30)$$

where we have normalized the result so that $R(0) = 1$. It is interesting to note that if we define the correlation time as

$$\tau_c = 1/f_{3dB} \qquad (8.31)$$

the autocorrelation function evaluated at τ_c is given by

$$R(\tau_c) = e^{-(\pi/\alpha)^2}$$
$$\simeq .03 \qquad (8.32)$$

The autocorrelation function drops off quickly. At $\tau = 1.2\tau_c$, $R(\tau) = .005$ and at $\tau = 1.5\tau_c$, $R(\tau) = .0003$. For our purposes it is reasonable to accept Eq. (8.31) as the definition of correlation time. In most cases it would make little difference in the results if we used a slightly different value for τ_c.

To generate random sequences we will begin with the general methods outlined in the previous section. The first thing to determine is the size of Δf, the spectral sample spacing. In Fig. 8.2 we have shown the relationship between the correlation time and the repetition period for the discrete process. Since τ_c is the correlation time and $1/\Delta f$ is the repetition period, the longest sequence we can use that does not encounter correlation from the repeated component is $N\Delta t$ where

$$N\Delta t \leqslant 1/\Delta f - \tau_c \qquad (8.33)$$

BEGIN WITH THE POWER SPECTRAL SEQUENCE $\{\hat{S}_n\}$. THE FILTER SEQUENCE IS GIVEN BY

$$\hat{H}_n = \sqrt{\hat{S}_n} \qquad , n = 0, \ldots, N_r - 1$$

TAKE THE DISCRETE FOURIER TRANSFORM TO OBTAIN THE IMPULSE RESPONSE $\{h_k\}$ AS

$$h_k = \sum_{n=0}^{N_r - 1} \hat{H}_n \, e^{j2\pi kn/N_r}, \quad k = 0, \ldots, N_r - 1$$

NOW GENERATE AN INDEPENDENT RANDOM PHASOR SEQUENCE $\{\hat{\zeta}_k\}$ AS

$$\{\hat{\zeta}_k\} \qquad\qquad \overline{|\hat{\zeta}_k|^2} = \frac{1}{N_r}$$

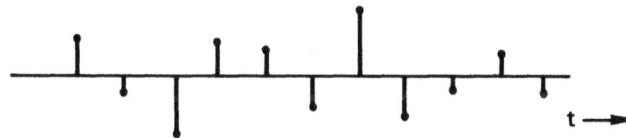

THEN PERFORM A DISCRETE CONVOLUTION TO OBTAIN DESIRED TIME-DOMAIN SEQUENCE $\{x_k\}$ AS

$$x_k = \hat{\zeta}_k \bigstar h_k$$

Figure 8.3/Alternate Approach to Generate Correlated Time Sequence (valid primarily for Gaussian process).

If we solve for Δf we can write

$$\Delta f \leqslant \frac{1}{\tau_c + N\Delta t} \qquad (8.34)$$

It is helpful to relate Δf to f_{3dB}. Thus we can write

$$\frac{f_{3dB}}{\Delta f} \geqslant \tau_c f_{3dB} + Nf_{3dB}\Delta t \qquad (8.35)$$

But from Eq. (8.31), $\tau_c f_{3dB} = 1$. Furthermore, we can set $\Delta t = 1/f_r$. Thus

$$\frac{f_{3dB}}{\Delta f} \geqslant 1 + N(f_{3dB}/f_r) \qquad (8.36)$$

This expression is convenient to use. We have a simple solution for the number of samples within the half-power width of the spectrum. We will now investigate two cases.

Case 1 — $N \lesssim f_r/f_{3dB}$

For the case where N is small enough so that $N \lesssim f_r/f_{3dB}$, we find that the right-hand side of Eq. (8.36) is somewhere between 1.0 and 2.0. This fact means that we will need, at most, only 2 samples within the half-power width of the spectrum. Furthermore, since the Gaussian function in Eq. (8.28) rapidly goes to zero for $|f| > f_{3dB}$, we may need as few as 5 or 7 spectral samples to accurately approximate the power spectral density. If we were to use the general expression in Eq. (8.22) to transform the random spectral sequence to a correlated time sequence, most of the spectral samples would be nearly zero. While Eq. (8.22) can be implemented with Fast Fourier Transform algorithms, it will usually be more efficient to use a brute-force transform on only those samples that are nonzero.

To construct the power spectral sequence $\{\hat{S}_n\}$ we will sample the power spectral density over M samples as

$$\hat{S}_n = \Delta f \cdot S\left[\left(n - \frac{M+1}{2}\right)\Delta f\right], n=1, 2, \dots, M \qquad (8.37)$$

For example: if M = 5, the samples will occur at f = -2Δf, -Δf, 0, Δf, and 2Δf. Note that the

samples will be symmetrical about the origin. To generate the correlated time sequence we first create the sequence of independent random phasors $\{\xi_n\}$, each of unit average power. Then we form

$$\hat{X}_n = \xi_n \sqrt{\hat{S}_n}, \qquad n=1,\dots,M \qquad (8.38)$$

Note that the ensemble average power of the process is

$$P = \sum_{n=1}^{M} \overline{|\hat{X}_n|^2} = \sum_{n=1}^{M} \hat{S}_n \qquad (8.39)$$

In general the Gaussian-shaped spectrum will not be centered at dc. For the case where it is offset from dc by f_o we can generate the correlated time sequence by implementing a discrete Fourier transform on Eq. (8.38) as

$$x_k = \sum_{n=1}^{M} \hat{X}_n e^{j2\pi[f_o + (n-\frac{M+1}{2})\Delta f]k\Delta t} \qquad (8.40)$$

Now we have to establish what the best choices for Δf and M are.

In choosing an optimum sample spacing Δf and the number of spectral samples M, one approach would be to compare the total power of the discrete sequences with the original continuous spectrum. In other words we can compare

$$\sum_{n=1}^{M} \hat{S}_n \quad \text{and} \quad \int_{-\infty}^{\infty} S(f)df$$

However, in radar simulations it is more appropriate to compare the results after the noise spectrum has been filtered. In Fig. 8.4 we show three general filter shapes defined by

$$H(f) = f^{2m}$$

for m = 0, 1, and 2. Now we can compare the filtered residues

$$\sum_{n=1}^{M} [(n-\frac{M+1}{2})\Delta f]^{2m} \hat{S}_n \quad \text{and} \quad \int_{-\infty}^{\infty} f^{2m}S(f)df$$

Note that the m = 0 case is also representative of the case where the filter notch is not centered on the Gaussian spectrum. In Fig. 8.5 we show the relative errors (relative to the filtered residue) associated with a given choice of Δf and M. From this figure we can readily see that the relative error is reduced as we increase M, provided we choose Δf properly. Just what error is tolerable? This question ultimately depends on the particular application. In radar simulations we will generally be able to tolerate errors of a few percent, as we have argued previously. While we can achieve this error with M = 3 or 4 under certain special conditions, we are not able to satisfy the "few percent error" criterion for all three filters until we have M = 5 or larger. For M = 5, the optimum value of Δf is approximately 0.60 f_{3dB}. We can reduce the errors somewhat by increasing M, but the increased accuracy is probably not worth the added computation. This consideration might be important, however, when the spectral shape is no longer Gaussian. The relative insensitivity of the error to Δf for M = 6 and 7 might make these values attractive for non-Gaussian spectral shapes.

In Fig. 8.6 we summarize the above procedure for generating a correlated time sequence with a Gaussian-shaped spectrum. In this example we have used $\Delta f = 0.6\ f_{3dB}$ and M = 5.

Case 2 – N > f_r/f_{3dB}

If we go back to Eq. (8.36) for the case where N > f_r/f_{3dB}, we find that

$$\frac{f_{3dB}}{\Delta f} \geqslant N \tag{8.41}$$

Thus there must be at least N samples within the half-power width of the Gaussian-shaped spectrum. In this situation we will find no obvious best approach to generating the correlated time sequence.

If we implement Eq. (8.22) directly by means of a Fast Fourier Transform (FFT), we must choose N_r as

$$N_r = \frac{f_r}{\Delta f} = \frac{f_r}{f_{3dB}}\frac{f_{3dB}}{\Delta f} \tag{8.42}$$

But the second factor on the right must satisfy Eq. (8.41), so that

$$N_r \geqslant N\ \frac{f_r}{f_{3dB}} \tag{8.43}$$

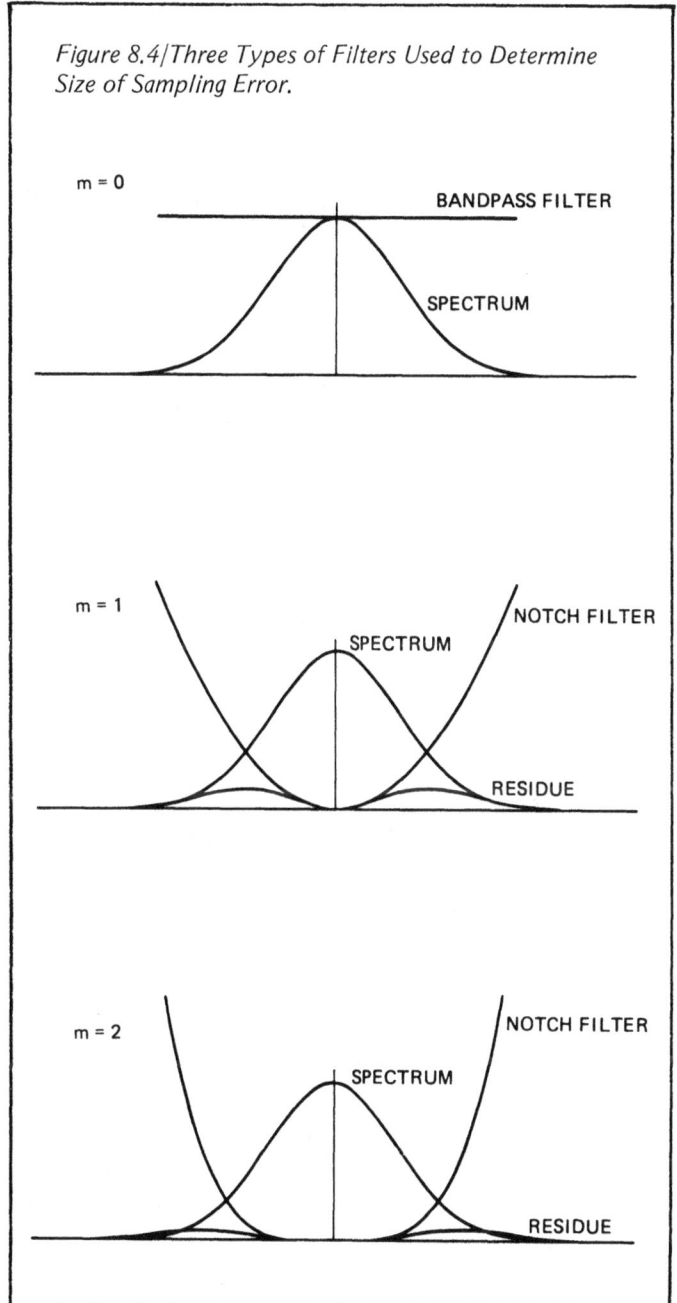

Figure 8.4/Three Types of Filters Used to Determine Size of Sampling Error.

Figure 8.5/Relative Errors Associated with Sample Spacing Δf and Number of Samples M.

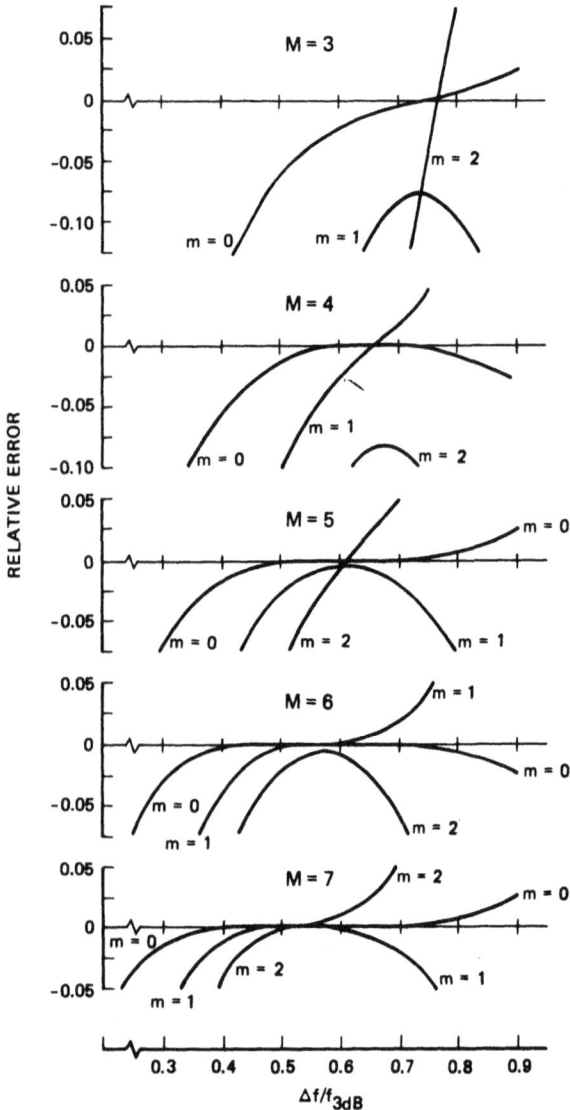

If we implement a brute-force approach as we did for Case 1, the number of spectral samples chosen depends on where we truncate the Gaussian function. If we truncate at twice the half-power width, we require at least 2N samples. As long as $N \gg f_r/f_{3dB}$, the comparison in running time between the FFT approach and

the brute-force approach will almost always favor the former. It is only when N is slightly larger than f_r/f_{3dB}, say within about a factor of 2, that the brute-force approach might be faster (the comparison, of course, depends on the particular computer and algorithms used). There is a third alternative for Case 2, the so-called time-domain approach based upon the convolution in Eq. (8.27), although there are certain limitations on its use. The first is the one discussed in Section 8.3 regarding Gaussian processes — only for the Gaussian process (and certain transformations thereof) can we begin with an independent Gaussian phasor sequence in the time domain before we perform the convolution. In Eq. (8.31) we show that the correlation time τ_c can be defined as the inverse half-power width of the Gaussian-shaped spectrum. Thus we can write

$$\frac{\tau_c}{\Delta t} = \frac{1}{f_{3dB}\Delta t} = \frac{f_r}{f_{3dB}} \qquad (8.44)$$

The left-hand side is the maximum number of time samples that are correlated with more than 3% correlation. But $f_r/f_{3dB} < N$ as a condition of Case 2. Let us go back to Eq. (8.27) which we copy here as

$$x_k = h_k \star \hat{\zeta}_k$$

where $\{h_k\}$ is the filter sequence operating on the independent (and probably Gaussian) random phasor sequence $\{\hat{\zeta}_k\}$. If K is the maximum number of samples in $\{x_k\}$ that are correlated, then the length of $\{h_k\}$ need be no longer than K. We will postpone until later the question of whether Eq. (8.44) is a satisfactory value for K. We will assume for the moment that K is approximately the value given by Eq. (8.44) and, as a condition of Case 2, $K < N$.

To generate a sequence $\{x_k\}$ of length N we can begin with the sequence $\{\hat{\zeta}_n\}$ of length N+K−1 and perform the discrete convolution

$$x_k = \sum_{m=1}^{K} h_m \, \hat{\zeta}_{k+K-m} \, , \qquad k=1,\ldots,N \qquad (8.45)$$

Figure 8.6/Procedure for Generating Correlated Random Sequence with Gaussian-Shaped Spectrum, Case 1: Frequency-Domain Approach.

Case 1: $N \lesssim f_r/f_{3dB}$ Frequency-Domain Approach

Given: P = ensemble average power of random process

f_{3dB} = half-power width of Gaussian-shaped spectrum

f_o = center frequency

Δt = sample spacing in time domain ($f_r = 1/\Delta t$)

N = number of samples of random process

Choose: $\Delta f = 0.6 \, f_{3dB}$

$M = 5$

The 5 spectral components are

n	1	2	3	4	5
$\hat{S}_n/P$	.0104	.2078	.5637	.2078	.0104
$\sqrt{\hat{S}_n/P}$	.1020	.4558	.7508	.4558	.1020

Compute only once $\sqrt{\hat{S}_n}$, n=1, ..., 5

Generate 5 uncorrelated random phasors ξ_n, n=1, ..., 5, each of unit average power.

Form $\hat{X}_n = \xi_n \sqrt{\hat{S}_n}$, n=1, ..., 5

Compute $x_k = \displaystyle\sum_{n=1}^{5} \hat{X}_n \, e^{j2\pi[f_o+(n-\frac{M+1}{2})\Delta f]k\Delta t}$, k=1, ..., N

---

where we have adopted the convention that all sequences begin with the index of unity.

The filter sequence is easily obtained if the spectrum is Gaussian shaped. From Eq. (8.28) and the fact that $H(f) = \sqrt{S(f)}$, we can write the Fourier transform pair of $H(f)$ as

$$h(t) = e^{-2(\pi f_{3dB}t/\alpha)^2} \tag{8.46}$$

where we have neglected a constant scale factor. This expression applies to a spectrum centered at dc. If the spectrum is centered at f_o, then

$$h(t) = e^{-2(\pi f_{3dB}t/\alpha)^2} \, e^{j2\pi f_o t} \tag{8.47}$$

If we sample at increments of Δt, we can write

$$h_k = e^{-7.1194(f_{3dB}/f_r)^2 (k-\frac{K+1}{2})^2} \, e^{j2\pi f_o(k-\frac{K+1}{2})\Delta t} \tag{8.48}$$

$$k=1, \ldots, K$$

Since we have neglected the scale factor in deriving Eq. (8.46), we must relate the sequence $\{h_k\}$ to the ensemble average power in the process $\{x_k\}$. If each sample in the random phasor sequence $\{\hat{\xi}_k\}$ is of unit average power, the ensemble average power in $\{x_k\}$ is easily derived from Eq. (8.45) as

$$P = \overline{|x_k|^2} = \sum_{k=1}^{K} |h_k|^2 \tag{8.49}$$

We will use this result to normalize the filter samples.

The one remaining question is how large should K be. To decide this point, we will compute the power spectrum associated with the filter sequence $\{h_k\}$; if it is close to the assumed Gaussian-shaped spectrum, then the procedure is valid. In Fig. 8.7 we show several power spectra for the case of $f_r/f_{3dB} = 8$. If we use Eq. (8.44) to set K, then K would also be 8. However, we see that the choice of K = 8 in Fig. 8.7 results in peak spectral sidelobes that are 29 dB below the mainlobe peak. The situation is improved if we go to higher values of K (which we tabulate below as the product $K(f_{3dB}/f_r)$; the results are based on the case of $f_r/f_{3dB} = 8$, but are assumed to apply for any value larger than unity):

$K(f_{3dB}/f_r)$	Peak Sidelobe Level
1.000	-29 dB
1.125	-33
1.250	-39
1.375	-44
1.500	-50
1.625	-57
1.750	-64

For example to get the peak sidelobes down to -50 dB, we must have K = 12 or 1.50 (f_r/f_{3dB}). In Fig. 8.8 we summarize the above procedure for the particular choice of K = 1.25 (f_r/f_{3dB}) designed to give peak sidelobes of -39 dB.

The filter $\{h_k\}$ obtained above is nonrecursive. If K is large, there might be a possibility of designing a recursive filter that would require less computation to implement. We can use conventional digital filter design techniques, utilizing such tools as the Z-transform, to create these filters. However, it will still be difficult to design recursive filters with extremely low spectral sidelobes that also will be as fast as the nonrecursive filters discussed above.

A discussion of rapid generation of random sequences would not be complete without some mention of the *Markov process*, a simple recursive filtering of independent samples. In simulations where an accurate rendition of the correlation properties of a random sequence is not important, the Markov process, especially one of first order, provides an extremely simple means of generating correlated random samples.

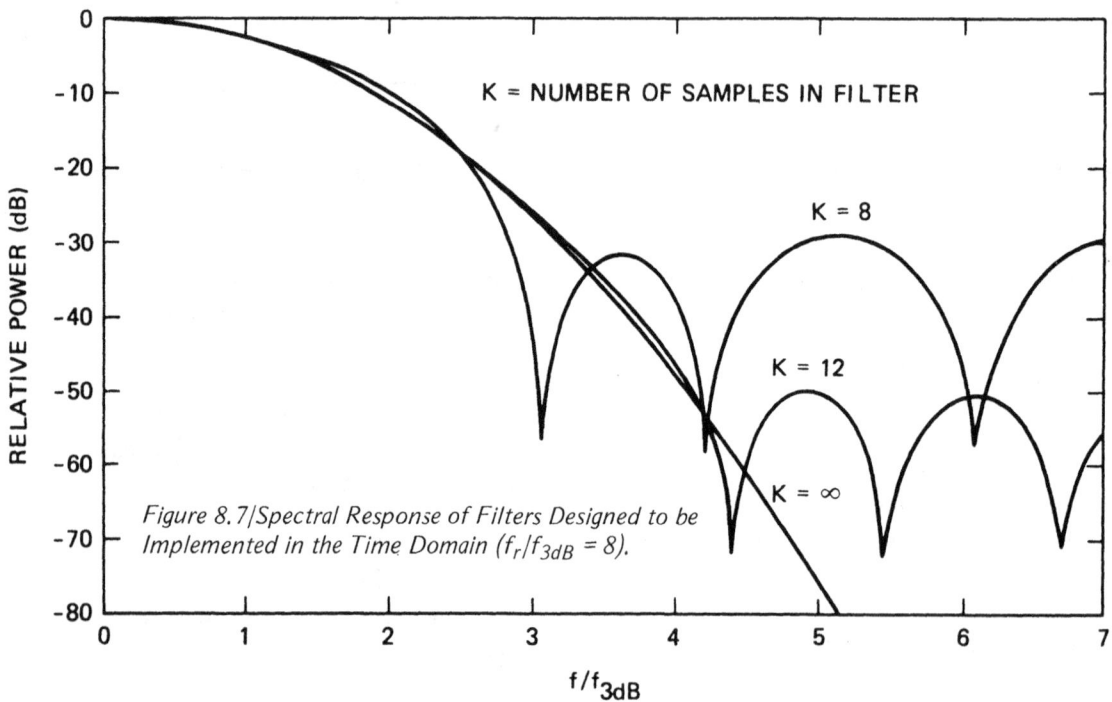

Figure 8.7/Spectral Response of Filters Designed to be Implemented in the Time Domain ($f_r/f_{3dB} = 8$).

Figure 8.8/Procedure for Generating Correlated Random Sequence with Gaussian-Shaped Spectrum, Case 2: Time-Domain Approach (procedure is generally limited to Gaussian processes).

Case 2: $N > f_r/f_{3dB}$ Time-Domain Approach

Given:
P = ensemble average power of random process
f_{3dB} = half-power width of Gaussian-shaped spectrum
f_o = center frequency
Δt = sample spacing in time domain ($f_r = 1/\Delta t$)
N = number of samples of random process

Choose: $K = 1.25\, f_r/f_{3dB}$ (spectral sidelobes 39 dB down)

Compute only once:
$$h_k = e^{-7.103(f_{3dB}/f_r)^2 (k-\frac{K+1}{2})^2}\, e^{j2\pi f_o(k-\frac{K+1}{2})\Delta t}$$

$$k = 1, \ldots, K$$

Normalize so that: $P = \sum_{k=1}^{K} |h_k|^2$

Generate $N+K-1$ uncorrelated random phasors $\hat{\zeta}_k$, $k=1, \ldots, N+K-1$, each of unit average power

Compute $x_k = \sum_{m=1}^{K} h_m\, \hat{\zeta}_{k+K-m}$, $k=1, \ldots, N$

Given a sequence of complex independent samples $\{\zeta_k\}$ of zero mean and unit average power we can generate a Markov sequence of first order with the recursive algorithm

$$x_k = \rho_1\, x_{k-1} + b\zeta_k \qquad (8.50)$$

where $\rho_1 = R_1/R_o = R_1/P$ is the normalized autocorrelation function of the sequence $\{x_k\}$ at one unit of lag; it can be arbitrarily specified, as well as complex. The parameter b is a scalar and is given by

$$b^2 = P[1-|\rho_1|^2] \qquad (8.51)$$

where P is the desired average power of the sequence $\{x_k\}$. The autocorrelation function of this sequence is determined explicitly by ρ_1 as

$$\rho_n = R_n/P = \rho_1^n \qquad (8.52)$$

There are higher-order Markov sequences that can be generated. Unfortunately the applicable processes are limited in the same way as are the general recursive filters; it is difficult to design sequences with extremely low spectral sidelobes.

8.5/Utilizing Interpolation to Reduce Sampling Frequency

If the ratio f_r/f_{3dB} is large for compact spectra such as those discussed in the last section, many consecutive samples in the time domain will be highly correlated, as we can see from Eq. (8.42). If we use any of the methods discussed in the last two sections to generate the correlated time sequence, there will be many computations to perform, unless of course N is small. Since little is changing from one time sample to another, an alternate method is to generate less frequent time samples in the conventional manner and use interpolation in between. This technique will be especially valuable in real-time simulations, as we will show in Chapter 12.

To illustrate the consequences of interpolation let us generate a random sequence $\{x'_k\}$ with samples spaced by $\Delta t'$ where $\Delta t' \gg \Delta t$, the desired sample spacing after interpolation. For the moment we constrain the ratio $\Delta t'/\Delta t$ to be an integer so the number of samples that we interpolate between each of the samples spaced by $\Delta t'$ is given by

$$N_I = \frac{\Delta t'}{\Delta t} = \frac{f_r}{f'_r} \qquad (8.53)$$

where $f_r = 1/\Delta t$, the desired repetition frequency (strictly speaking the number of interpolated samples will be $N_I - 1$). In effect $f'_r = 1/\Delta t'$ is the new repetition frequency. We show such a sequence in Fig. 8.9. To create the samples in between those in $\{x'_k\}$ we will interpolate, which

is equivalent to convolving the sequence $\{x'_k\}$ with the interpolation function i(t). For linear interpolation

$$i_1(t) = 1 - |t/\Delta T|, \ |t| \leqslant \Delta T$$
$$\qquad\qquad\qquad\qquad\qquad (8.54)$$
$$\quad = 0 \qquad\qquad , \text{ otherwise}$$

Convolving in the time domain is equivalent to multiplying spectra in the frequency domain. Thus if $X'(f)$ is the repeated spectrum corresponding to the time sequence $\{x'_k\}$ and $I(f)$ is the Fourier transform of i(t), the spectral product is given by

$$X''(f) = X'(f)I(f) \qquad (8.55)$$

In Fig. 8.9 we have sketched the result. Note the existence of residual sidelobes spaced at multiples of $f'_r = 1/\Delta t'$. This spectral function must be compared to the desired function that vanished everywhere except at the repetition frequencies spaced by $f_r = 1/\Delta t$.

The interpolation function spectrum is simple to express for the various interpolation formulas. The principle ones of interest are:

Sample & Hold (1-point)	$I_1(f) = \text{sinc}(f\Delta t')$
Linear (2-point)	$I_2(f) = \text{sinc}^2(f\Delta t')$
Quadratic (3-point)	$I_3(f) = \text{sinc}^3(f\Delta t')$
General (n-point)	$I_n(f) = \text{sinc}^n(f\Delta t')$

$$(8.56)$$

where we have included the simple case of sample and hold for comparison. While the interpolation is implemented in the time doamin, the effect of interpolation is best analyzed in the frequency domain. The residual sidelobes spaced at multiples of $f'_r = 1/\Delta t'$ are not desired, since no sidelobes would be present if we generated the sequence directly at a sample spacing of Δt. The residual power associated with one of the sidelobes is given by

$$R_n = \int |X''(f)|^2 df = \int |X'(f)|^2 \text{sinc}^{2n}(f\Delta t') df \qquad (8.57)$$

where the integration is performed in the neighborhood of the sidelobe. Note that in this region

Figure 8.9/Method of Utilizing Interpolation to Reduce Sampling Frequency.

THE GOAL IS TO GENERATE A CORRELATED SEQUENCE $\{x_k\}$ SPACED BY $\Delta t = 1/f_r$ AS

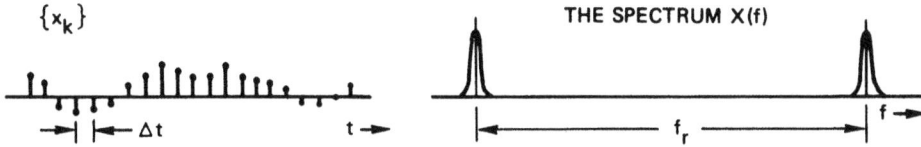

$\{x_k\}$ — THE SPECTRUM X(f)

TO IMPLEMENT THE INTERPOLATION METHOD, FIRST GENERATE A RANDOM SEQUENCE $\{x'_k\}$ SPACED BY $\Delta t' = 1/f'_r$ WITH $\Delta t'/\Delta t$ EQUAL TO AN INTEGER AS

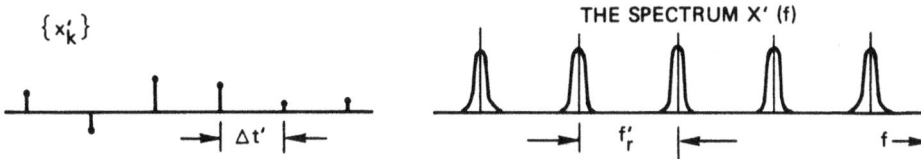

$\{x'_k\}$ — THE SPECTRUM X' (f)

TO CREATE THE SAMPLES IN BETWEEN THOSE IN $\{x'_k\}$, CONVOLVE $\{x'_k\}$ WITH AN INTERPOLATION FUNCTION i(t), SUCH AS THE ONE FOR LINEAR INTERPOLATION GIVEN BY

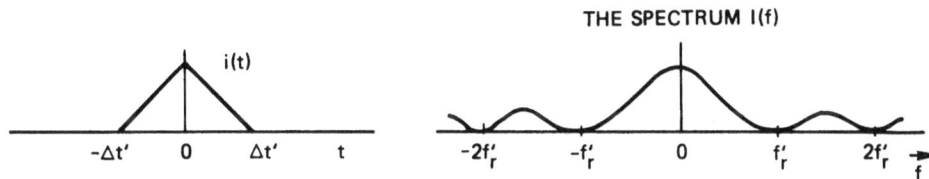

i(t) — THE SPECTRUM I(f)

THE RESULT OF THE CONVOLUTION AND THE SPECTRAL PRODUCT IS

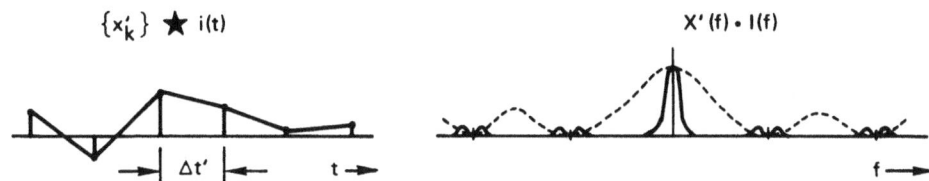

$\{x'_k\}$ ★ i(t) — X' (f) · I(f)

NOW SAMPLE THE RESULTING TIME FUNCTION AT INTERVALS OF Δt AS

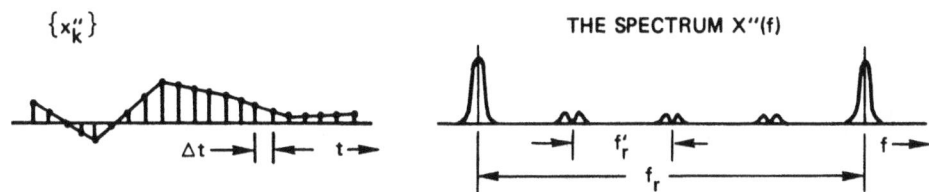

$\{x''_k\}$ — THE SPECTRUM X''(f)

Figure 8.10/Residual Power in Spectral Sidelobe Caused
by Interpolation.

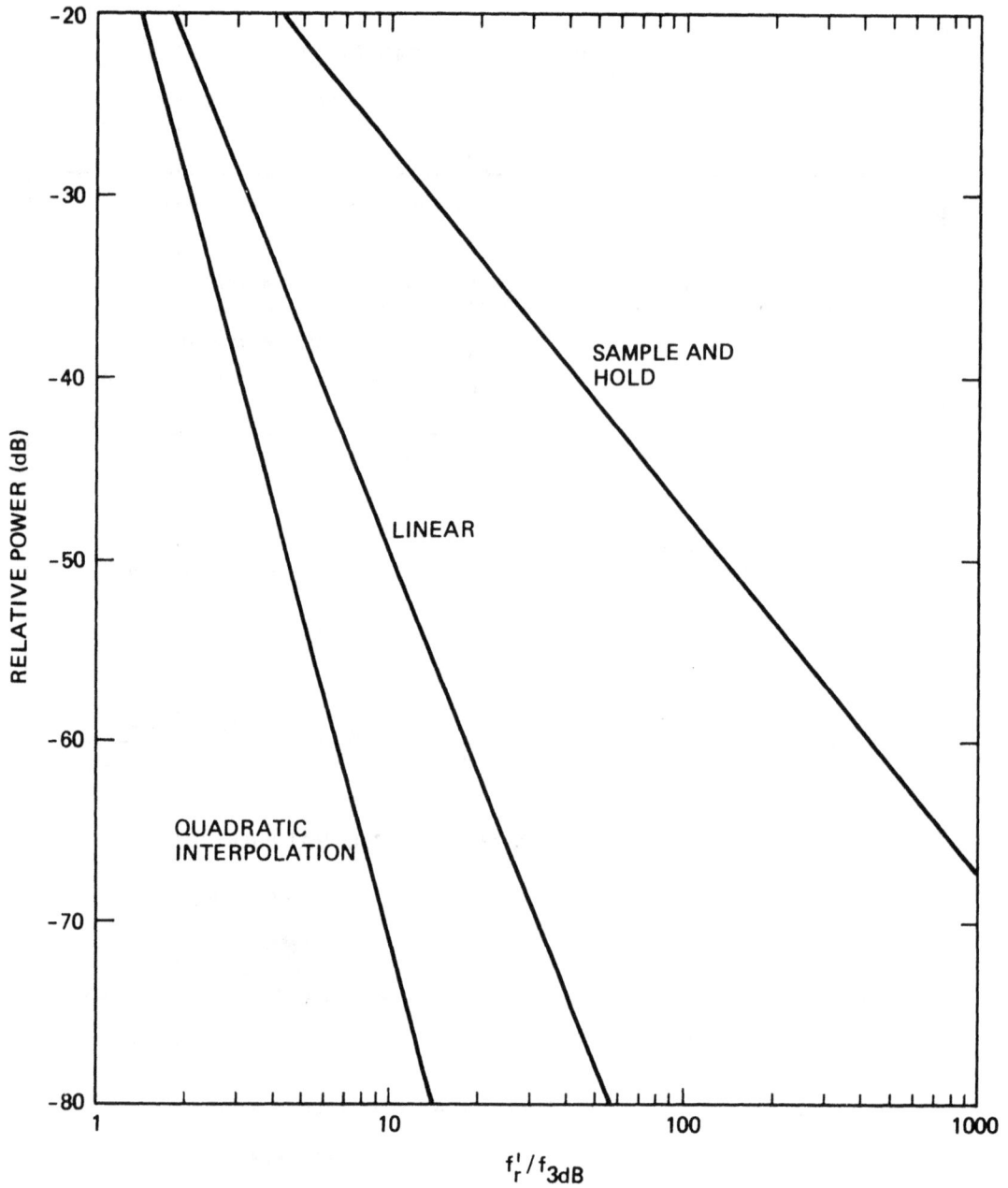

$$|\text{sinc}(m+f\Delta t')| = \left|\frac{f\Delta t'}{m}\right| \quad (8.58)$$

for f small where m is any integer except zero. Thus as long as the extent of the spectral component in $X'(f)$ is small compared to $1/\Delta t'$ (i.e., $f_{3dB}\Delta t' \ll 1$), we can write the largest component of residual power, namely that appearing in the first sidelobe (m = 1), as

$$R_n = \int |X'(f)|^2 (f\Delta t')^{2n} df \quad (8.59)$$

Now for the Gaussian-shaped spectrum we can readily perform the integration over one component of $X'(f)$, assuming infinite limits, to obtain

$$R_n = \frac{1\cdot3\cdot5\dots(2n-1)}{2^n}\left(\frac{f_{3dB}\Delta t'}{\alpha}\right)^{2n} \quad (8.60)$$

where $\alpha = 1.665$ and f_{3dB} is the half-power width of the Gaussian-shaped spectrum. The result in Eq. (8.60) is normalized to the power in the central lobe, which is given by Eq. (8.59) for n = 0 (i.e., $R_0 = 1$). The principal values are:

Sample & Hold (n=1) $R_1 = 0.180(f_{3dB}\Delta t')^2$

Linear (n=2) $R_2 = 0.098(f_{3dB}\Delta t')^4$ **(8.61)**

Quadratic (n=3) $R_3 = 0.088(f_{3dB}\Delta t')^6$

Since $f_{3dB}\Delta t' \ll 1$, the residual powers in Eq. (8.61) are indeed small. We show these results in Fig. 8.10 as a function of $f_r'/f_{3dB} = 1/f_{3dB}\Delta t'$.

The relative residual power that is acceptable depends primarily on the Doppler filter sidelobe levels designed in the radar system. If this level is -42 dB, as it would be with Hamming weighting, we should strive to keep the residual power due to interpolation about 5 dB lower, or about -47 dB. This limiting would mean

(n=1) $f_r' \geqslant 95 f_{3dB}$

(n=2) $f_r' \geqslant 8.36 f_{3dB}$

(n=3) $f_r' \geqslant 4.05 f_{3dB}$

Now the number of samples that we can interpolate between the samples spaced by $\Delta t'$ is

given by Eq. (8.53). If we substitute the above values for f_r' into Eq. (8.53), we obtain

(n=1) $N_I \leqslant .0105 f_r/f_{3dB}$

(n=2) $N_I \leqslant .120 f_r/f_{3dB}$

(n=3) $N_I \leqslant .247 f_r/f_{3dB}$

Thus if $f_r/f_{3dB} = 100$, we have a maximum number of interpolated points as

(n=1) $N_I = 1$

(n=2) $N_I = 12$

(n=3) $N_I = 24$

The first result is that sample and hold does not buy us anything for this example if we want to keep the residual sidelobes below -47 dB. A tradeoff can now be made between linear interpolation (n=2) over 12 consecutive samples and quadratic interpolation over 24 consecutive samples. Although there will be only half of the computation time involved in generating the random samples for the latter case (only half as many samples will be generated in a given time interval), quadratic interpolation requires more effort than linear interpolation.

When we began this discussion, we constrained the ratio $\Delta t'/\Delta t = f_r/f_r'$ to be an integer; actually, this condition is not necessary. The only requirement is that f_r' be chosen large enough so that the spurious spectral sidelobes are kept below some specified value. If we visualize the interpolation function i(t) as being continuous, we can resample the continuous random function $\{x_k'\} \star i(t)$ at any rate, even one that is less than f_r'. However, unless f_r/f_r' is an integer, we will have to recompute the interpolation weights for each new sample.

Reference

[1] Broste, N.A.; "Digital Generation of Random Sequences with Specified Autocorrelation and Probability Density Functions," *U.S. Army Missile Command Report RE-TR-70-5*, March 1970.

Chapter 9

Random Number Generators and Other Library Functions

In the preceding chapter we discussed several methods of generating random sequences; in this chapter we will show how the random numbers themselves can be generated. First we will describe some classical methods for generating random numbers that are of interest in radar simulation. Then we will give some fast algorithms that greatly reduce the computation time, including some algorithms for standard library functions. For large scale simulations and for those that run in real-time, it is imperative that such techniques are used to reduce the number of computations.

9.1/Classical Methods for Generating Random Numbers

There are many methods used to generate random numbers [Ref. 1]. Some are more efficient than others for a particular type of random number. In this section we will discuss certain classical methods which result in an essentially continuously distributed variate. Examples are given only for the distribution functions that are popular in radar simulations.

Components of Unit-Amplitude Random Phasor

In some cases it is expedient to generate the amplitude and phase of a random phasor separately. Thus if we generate a random phasor of unit amplitude, we can scale it by the random amplitude to create the desired random phasor. Let c and

s be the real and imaginary components of the random phasor of unit amplitude such that

$$c = \cos\theta$$
$$s = \sin\theta \qquad (9.1)$$

and θ is a random angle distributed uniformly over $(0, 2\pi)$. Note that $c^2 + s^2 = 1$ and the components c and s are independent. To generate a random phasor $c + js$, we can generate a random value of θ and perform the sine and cosine operations in Eq. (9.1).

We can avoid the sine and cosine computation if we employ an algorithm first developed by von Neumann. Let us write two identities as

$$c = \cos\theta = \cos^2\frac{\theta}{2} - \sin^2\frac{\theta}{2} \qquad (9.2)$$
$$s = \sin\theta = 2\sin\frac{\theta}{2}\cos\frac{\theta}{2}$$

Now define A and B as sides of a right triangle, as we show in Fig. 9.1, such that

$$\cos\frac{\theta}{2} = \frac{A}{\sqrt{A^2 + B^2}}$$
$$\sin\frac{\theta}{2} = \frac{B}{\sqrt{A^2 + B^2}} \qquad (9.3)$$

If we substitute these identities into Eq. (9.2), we obtain

$$c = \cos\theta = \frac{A^2 - B^2}{A^2 + B^2} \qquad (9.4)$$
$$s = \sin\theta = \frac{2AB}{A^2 + B^2}$$

If A and B are given, we can now compute s and c by straightforward arithmetic operations. Now let us generate A and B as independent random samples from a uniform distribution over $(-1, 1)$ (e.g., if u is uniformly distributed over $(0, 1)$, then $A = 2u-1$). If A^2+B^2 happens to be greater than unity, we will continue to generate random pairs until A^2+B^2 falls below that level (a so-called acceptance-rejection technique). We can treat A and B as being distributed along orthogonal axes at the center of the unit circle in Fig. 9.1 so that we are in effect generating a coordinate A+jB that is uniformly distributed within the unit circle. Hence, the angle $\theta/2$ defined by Eq. (9.3) is uniformly distributed over $(0, 2\pi)$. It follows that θ modulo 2π is also uniformly distributed over the same interval.

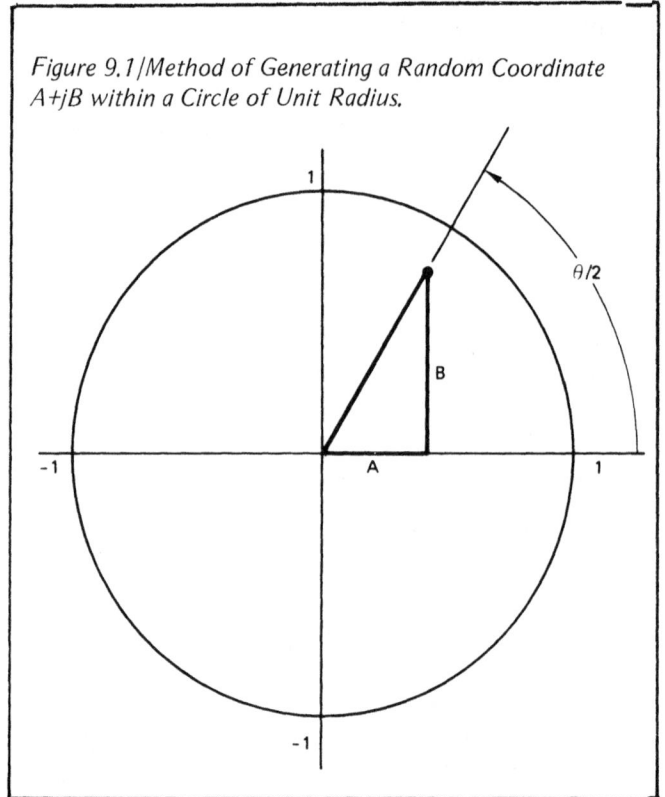

Figure 9.1/Method of Generating a Random Coordinate A+jB within a Circle of Unit Radius.

Exponential Distribution

The most common distribution of power is the exponential distribution. To generate a random sample ε from an exponential distribution of unit mean, set

$$\varepsilon = -\ln(u) \qquad (9.5)$$

where u is a random sample from a uniform distribution over $(0, 1)$.

Rayleigh Distribution

If power is exponentially distributed, the the amplitude or the square-root of power will be Rayleigh distributed. To generate a random sample r from a Rayleigh Distribution where the second moment is unity, set

$$r = \sqrt{\varepsilon} \qquad (9.6)$$

where ε is given by Eq. (9.5).

Gaussian Distribution

The so-called Gaussian process refers to the case where the components of a random phasor are normally or Gaussian distributed. For this case the amplitude of the phasor is Rayleigh (we are restricting the process to be of zero mean) and the power is exponentially distributed. Thus we can generate two independent Gaussian random samples g_1 and g_2 by setting

$$g_1 = r \cdot c$$
$$g_2 = r \cdot s \qquad (9.7)$$

where r is given by Eq. (9.6) and c and s by Eq. (9.1) or Eq. (9.4). We note that

$$\overline{g_1^2} + \overline{g_2^2} = 1 \qquad (9.8)$$

so that the variance of each component is 0.5 (half of the power goes into each component). To generate a Gaussian random sample of unit variance we must multiply g_1 or g_2 by $\sqrt{2}$.

There are other methods of generating Gaussian random samples. One popular method is based on adding 12 independent random uniform samples as

$$g = \sum_{k=1}^{12} u_k - 6 \qquad (9.9)$$

where u_k is uniformly distributed over $(0, 1)$. In this case the mean value of g will be zero and the variance unity.

Log-Normal Distribution

Another common distribution of power in radar simulation is the log-normal distribution. If ℓ is a random sample from a log-normal distribution, then $\log_e \ell$ is from a Gaussian distribution. We denote the standard deviation of $\log_e \ell$ as σ_L. To generate a random sample ℓ set

$$\ell = e^{(g\sigma_L - 0.5\sigma_L^2)} \qquad (9.10)$$

where g is a random sample from the Gaussian distribution of zero mean and unit variance. The last factor in Eq. (9.10) makes the mean value of ℓ unity.

Weibull Distribution

Still another, but less common distribution of power is the Weibull distribution. It is characterized by a slope-parameter a and is a transformation of the exponential distribution as

$$\omega = \mathcal{E}^a / \Gamma(a+1) \qquad (9.11)$$

where the last factor makes the mean value ω unity.

Note that with the log-normal and Weibull distributions we can generate the amplitude directly without taking a square root if we substitute $\sigma_L/2$ for σ_L and a/2 for a, respectively.

9.2/Fast Algorithms for Generating Random Numbers*

The computation time in radar simulations is often dominated by the random number generator for non-uniform distribution functions. In most cases we would be willing to incur some error in the generation of the random number in order to greatly speed up the simulation. An extremely fast method is to approximate the continuous distribution function by a discrete one. A table of random numbers is precomputed, where each one occurs with the same probability. To achieve the large time savings the table must be constructed in such a way that the address or entry to the table can be obtained by a simple masking operation, as was suggested by Aus and Korn [Ref. 3]. To demonstrate such an implementation, we begin with the inverse method of generating random numbers.

A classical method of generating random numbers is the *inverse method* [Refs. 1 and 4]. Given a cumulative distribution function $F(x)$ of the random variable x, the inverse method is based on being able to obtain a solution for x of $u = F(x)$ for a given value of u. We write this as

$$x = F^{-1}(u) \qquad (9.12)$$

*The material in this section is taken from Ref. 2, with permission of the IEEE.

In other words u is now the independent variable. To generate a random sample x we generate one sample u from a uniform distribution over (0, 1) and apply the above transformation. For example if x is exponentially distributed with a unit mean, the distribution function is given by

$$F(x) = 1-e^{-x} \qquad (9.13)$$

The inverse is given by

$$x = F^{-1}(u) = -\ln(1-u) \qquad (9.14)$$

Although $x = -\ln(u)$ is also a valid transformation as we have shown in Eq. (9.5), we will use Eq. (9.14) in order to have $x \to \infty$ as $u \to 1$. While it is not always possible to obtain an analytical expression for the inverse of a distribution function, we can still implement the method by deriving an approximate expression for the inverse. There are other methods for generating continuous random variables, as we have shown in Section 9.1, but we will show that even the best of these methods is much slower than one based on a table-lookup.

The table-lookup approach is simple to implement. We divide the interval (0, 1) into N equal increments and select a representative random sample (e.g., the median) from the distribution function for that increment. We follow this procedure for each increment, as we have sketched in Fig. 9.2; note that it can be done for any arbitrary distribution function. The sequence of samples $\{x_k\}$ is stored in a table; choosing a random entry is now easy. Since the point probability associated with each increment in u is 1/N, we could generate a random entry to the table, k (k = 0, ..., N–1), by computing the integer value of Nu, where u is a uniform random sample over (0, 1). Moreover, if we choose $N = 2^K$, where K is an integer, we can generate a random entry by masking K bits of u. Not only is the masking operation fast, but one uniform random sample can also provide several sequences of K bits. However, care must be exercised to ensure that all bits are independent.

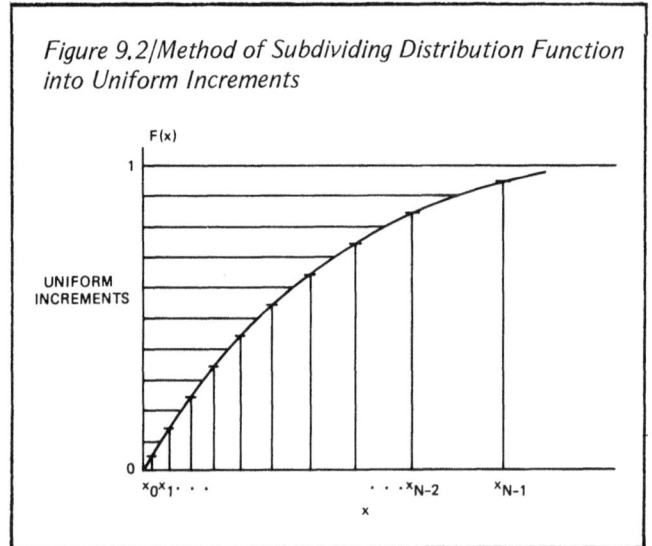

Figure 9.2/Method of Subdividing Distribution Function into Uniform Increments

(In many pseudo-random number generators the least significant bits have short periods.)

There are various ways to choose the representative sample of each increment — we could base it on the median value of u in the increment, which would be simple to compute, or we could choose a value that would minimize the mean square error of x in the increment (it is straightforward to show that this choice would be the mean value of x for the increment). However, the computation would no longer be simple. Of more importance is the fact that the resulting rms errors for the two choices usually differ only in the third significant decimal. For this reason it is recommended that the median values be used so that the table values are given by $x_k = F^{-1}(u_k)$ where

$$u_k = \frac{k + 1/2}{N}, \qquad k = 0, ..., N-1 \qquad (9.15)$$

For distribution functions where the domain is infinite, the largest errors occur in the last increment; the continuous distribution predicts occasional values that would exceed the last value in the table. A good way to alleviate this trucation-associated error is to apply the inverse method continuously over the last increment.

Some convenient approximation to the inverse function must be used if no tractable exact expression exists. Whenever the entry is to the last increment we would generate a random variable u, uniformly distributed over (1–1/N, 1), in order to use Eq. (9.12) to generate x over the last increment. Although there is more computational effort involved with this scheme whenever the entry to the last increment is selected, the situation occurs with a probability of only 1/N, which we can make small by making N large.*

Errors due to quantization can be reduced by linearly interpolating between discrete samples, but at a cost of increased computation time. To implement a linear interpolation scheme, the index k and the fraction are both random. If interpolation is used, the discrete sequence $\{x_k\}$ is constructed at the boundaries of the increments $u_k = k/N$, $k = 0, \ldots, N$. Let us first set

$$x_k = F^{-1}(k/N) \qquad (9.16)$$

for $k = 0, \ldots, N$, but as Aus and Korn [Ref. 3] show, there is a substantial mean error with such a procedure. Since the mean error is approximately 2/3 of the peak error in each increment, we can estimate the mean error by assuming that the peak error occurs at the median of the increment. Thus a good estimate of the amount to be subtracted from x_k in Eq. (9.16) to make the generator unbiased is

$$\varepsilon_k = \frac{2}{3} \left[\frac{x_{k+1} + x_k}{2} - F^{-1}\left(\frac{k + 1/2}{N}\right) \right] \qquad (9.17)$$

for $k = 1, \ldots, N-1$ (the end samples will not be adjusted). The rms error will then be reduced

to about 40% of the unbiased error (the subsequent error tabulations reflect this operation). Note that if the range of x is infinite, the last increment must be treated exactly by the inverse method (and the above bias should not be applied on the N–1st sample).

In Table 9.1 we show how the various table-lookup methods compare in terms of the errors and execution times for the exponential and Gaussian distribution functions. In the latter case we use Formula 26.2.22 in Ref. 1 for the inverse. Because of symmetry only the positive half of the Gaussian distribution function need be used; an additional random bit is selected for the sign of the random number. Three table-lookup methods are shown. They are:

Last Increment Discrete	– no compensation for last increment
Last Increment Exact	– inverse method over last increment
Linear Interpolation	– linear interpolation plus inverse method over last increment

To compare the table-lookup methods with the classical methods of generating random variables we note the following execution times for the CDC6400 in FORTRAN:

mulitply	5.5 μsec
Uniform (RANF)	22
Exponential, inverse method	150
Gaussian, inverse method	340
Gaussian, sum of 12 uniform	320
Gaussian, von Neumann algorithm [Ref. 4]	220

Thus the table-lookup methods are faster by a factor ranging from 5 to 20 in comparison to the more exact approaches.*

Before we select a particular method and table size, it is desirable to discuss just what error is acceptable. As we have already mentioned in

---

*Another method of reducing the error associated with truncation is to subdivide the last increment into finer, equally spaced increments and thus create a second-level sequence. Then if a given random entry is in the last increment of the first level, we generate a new random entry to the second-level table. We could extend this procedure to a third level, or higher. While the execution is somewhat faster than the exact method over the last increment, the additional storage required is not attractive.

---

*The comparison is even more favorable for the table-lookup methods on small computers without floating-point hardware.

Table 9.1/Errors and Execution Times for Exponential and Gaussian Distributions (the errors are relative to the standard deviation and the times are in μsec for assembly-language code on the CDC6400 and do not include any subroutine overhead)

N	Method	Exponential RMS Error	Time	Gaussian RMS Error	Time
32	last increment discrete	.184	8	.063	9
	last increment exact	.050	15	.022	21
	linear interpolation	.0048	25	.0012	31
64	last increment discrete	.130	8	.042	9
	last increment exact	.035	13	.013	16
	linear interpolation	.0034	23	.0008	26
128	last increment discrete	.092	8	.029	9
	last increment exact	.025	11	.010	14
	linear interpolation	.0024	21	.0005	24
256	last increment discrete	.065	8	.021	9
	last increment exact	.018	10	.007	12
	linear interpolation	.0017	20	.0004	22
512	last increment discrete	.046	8	.016	9
	last increment exact	.013	10	.005	12
	linear interpolation	.0012	20	.0002	22

Sections 6.4 and 8.4, radar simulations are generally based on some random phenomenon. Quantization errors introduced by these discrete approximations are one additional source of noise, but as long as the errors are small compared to the variability (i.e., standard deviation) of the random quantities, they will have a negligible effect on the overall simulation results. If quantization errors were additive and independent, the variance of the new discrete distribution would be the sum of the variance of the original distribution and the mean square quantization error. In this situation an rms quantization error of a few percent would be completely negligible. While such a conclusion is not rigorous, it seems justifiable to assume that quantization errors of one to three percent should be tolerable in most radar simulations.

Truncation errors are another matter, since the occasional events (e.g., false alarms) that are governed by the tails of a distribution function are affected considerably by quantization — that is, unless we use the inverse method over the last increment to avoid such truncation errors. Since the execution times are increased only slightly by this modification, it is highly desirable to implement the inverse method over the last increment in all applications where the domain of the distribution function is infinite.

To get back to Table 9.1 we see that we can achieve the one to three percent error with modest table sizes *without interpolation*. As a compromise between table size and execution time we recommend N = 128 for both distribution functions. Thus we can generate an exponential random variable in 11 μsec with a 2.5% rms error and a Gaussian random variable in 14 μsec with a 1% rms error.

It is straightforward to extend the method to other distribution functions. For example, to generate a Rayleigh distributed random variable, we would first construct a table based on the square-root of Eq. (9.14). Another example is the log-normal distribution. Here we must begin with a two-sided table for the Gaussian distribution function; each entry will then be transformed by Eq. (9.10) to create the log-normal table. The resulting rms errors are dependent on the choice of σ_L, and because of the long tail associated with the log-normal distribution it is imperative that the last increment be treated exactly by the inverse method. A table of length N = 128 with σ_L = 1 will result in an rms error of 4% (the error actually goes down as σ_L is increased because more of the area in the tail is being treated exactly). The execution time for this example is 13 μsec on the CDC6400. If interpolation is used, the rms error is reduced to 0.9% while the execution time is increased to 23 μsec. The continuous method of generating a log-normal random variable described in Section 9.1 requires 400 μsec on the CDC6400.

The components of a unit-amplitude random phasor are also easily generated by the table-lookup method. We need to construct only one table of cos θ for θ uniformly distributed over $(0, \pi/2)$. Since sin θ = cos$(\pi/2-\theta)$, the table entry for sin θ is the complement of the entry for cos θ. But we need two extra random bits to determine the signs of the phasor components since the table is constructed only in the first quadrant. The rms error for the discrete method without interpolation is about 0.45/N, and with interpolation it is about 0.091/N^2. With a table size of N = 32 and no interpolation the rms error is about 1.4%, which is adequate for most radar simulations. The execution time for this case is 13 μsec (per pair) on the CDC6400, which compares to 195 μsec for the von Neumann algorithm discussed in Section 9.1.

9.3/Fast Algorithms for Computing Library Functions

The library functions of square-root, log, exponential, sine, and cosine are usually computed with a precision that is compatible with the computer word length. In most radar simulations this high precision is not required and we can save a significant amount of computation time by using faster, but less accurate, algorithms. The general approach described in the last section is also well suited for the library functions.

The procedure we will use is straightforward. Let the continuous function be designated by y = y(x) where x is the independent variable for the argument of the subroutine. For all of the above library functions we will be able to construct a table to represent the function over a short interval of x, and values of the function outside of the limits of the table can be obtained by biasing or scaling the values within. Let us define a $\leqslant$ x $<$ b to be the interval over which the table is constructed, divided into N equal increments, where N is some integer power of 2 as N = 2^K (in order to be able to take advantage of the masking operation to find the table address). As before, the table values will be chosen at the median in each increment as

$$y_k = y(x_k) = y[a+(b-a)(2k+1)/2N] \qquad (9.18)$$

where $k = 0, \ldots, N-1$. During execution we will obtain the table entry k by masking some portion of the word containing x.

Errors due to quantization can be reduced by interpolating between discrete samples, but at a cost of increased computation time. If interpolation is used, we could compute the table values at the boundaries of the increments as

$$y_k = y(x_k) = y[a+k(b-a)/N] \tag{9.19}$$

where $k = 0, \ldots, N$ (we now need a table size of $N+1$). However, it is advantageous to correct the tables to remove the bias, as in Section 9.2, to reduce the resulting rms error. A satisfactory amount to subtract from the kth sample in the table is

$$\varepsilon_k = \frac{2}{3}\left[\frac{y_k + y_{k+1}}{2} - y\left(x_k + \frac{b-a}{2N}\right)\right] \tag{9.20}$$

for $k = 1, \ldots, N-1$ (the subsequent error specifications reflect this operation). During execution we will still obtain an index k, $0 \leqslant k \leqslant N-1$, by masking some part of x, but we will also obtain a fraction h, $0 \leqslant h < 1$, by masking a remaining part of x so that we can compute the approximate value of y as

$$\tilde{y} = y_k + h(y_{k+1} - y_k) \tag{9.21}$$

We can also store the differences $(y_{k+1} - y_k)$ in a second table to shorten the computation somewhat.

In deriving the discrete algorithms we will make use of both fixed-point and floating-point number formats. In fixed-point we will write any number z as

$$z = In(z) + Fr(z) \tag{9.22}$$

where $In(z)$ denotes the integer part of z and $Fr(z)$ denotes the fractional part of z. Similarly, in floating-point the number z will be written as

$$z = Co(z) \cdot 2^{Ex(z)} \tag{9.23}$$

where $Co(z)$ is the coefficient or mantissa of z and $Ex(z)$ is the exponent, which is an integer. The floating-point number will always be normalized so that $1 \leqslant Co(z) < 2$, thus placing the binary point to the right of the most significant bit of the coefficient (and that bit is always unity).

The address or entry to the table during execution will be obtained by masking the K most significant bits to the right of the binary point of either $Co(x)$ or $Fr(x)$, the independent variable, depending on the format of x. If interpolation is used, the remaining part of the word is converted to the fraction h used in Eq. (9.21).

We will now give the specific algorithms. The CDC6400 execution times quoted for each algorithm are for assembly-language code and do not include any check on the independent variable for invalid values; nor is the overhead associated with the subroutine call included. The errors given for the algorithms with interpolation include the bias used to reduce the mean error.

Square Root — $y = \sqrt{x}$

For the square-root algorithm let us write

$$y = \sqrt{x} = \sqrt{Co(x)} \cdot 2^{Ex(x)/2} \tag{9.24}$$

Let us also treat y as a floating-point number. Since the exponent must be an integer, we must treat two cases as

$$Ex(x) - even - \quad Co(y) = \sqrt{Co(x)}, \, 1 \leqslant Co(y) < \sqrt{2} \tag{9.24a}$$

$$Ex(y) = Ex(x)/2 \tag{9.24b}$$

$$Ex(x) - odd - \quad Co(y) = \sqrt{2Co(x)}, \, \sqrt{2} \leqslant Co(y) < 2 \tag{9.24c}$$

$$Ex(y) = [Ex(x)-1]/2 \tag{9.24d}$$

We will approximate $Co(y)$ in Eqs. (9.24a) and (9.24c) by constructing two discrete tables. Which table is used depends on the least significant bit of $Ex(x)$. The address or table entry during execution is determined by the K most significant bits to the right of the binary point of Co(x). The division by 2 in Eqs. (9.24b) and (9.24d) is accomplished by a right shift. The execution time for the CDC6400 is 13 μsec without interpolation and 23 μsec with interpolation, in comparison to the system library function SQRT which requires 86 μsec. The rms error relative to the interval $1 \leqslant \sqrt{x} < \sqrt{2}$ is about $0.12/N$ without interpolation and $0.0068/N^2$ with interpolation. A table size of N = 16 results in an rms error of less than 1% without interpolation.

We can greatly improve the accuracy of the square-root computation by using Newton's iterative method, which is given by

$$y_{new} = \frac{1}{2}\left[\frac{x}{y_{old}} + y_{old}\right] \qquad (9.25)$$

After each iteration the relative error is given by 1/2 of the square of the previous error. If we begin with a discrete approximation to y with a small table size and no interpolation, it is possible to obtain a very accurate result in a few iterations. This time let us work with the relative *peak* error, which is $0.25/N$ prior to the iteration. In Table 9.2, we show the relative peak error after various stages.

The inherent precision of the CDC floating-point word is about 10^{-14}. Therefore, with N = 8 and 3 iterations, we can achieve the same accuracy as the system library function, yet it requires only 40 μsec — less than half of the execution time. Note that we can get better than 10^{-6} accuracy with the same table of N = 8 with only 2 iterations.

Logarithm — y = log₂x

Base-2 logarithms are the logical choice for constructing algorithms. We can multiply the base-2 logarithm by a constant to convert the result to another base as $\log_b x = \log_2 x \cdot \log_b 2$. For the base-2 logarithm let us express x in floating point and y in fixed point. Then we can write

$$y = \log_2 x = \log_2[Co(x)] + Ex(x) \qquad (9.26)$$

where

$$In(y) = Ex(x) \qquad (9.26a)$$

$$Fr(y) = \log_2[Co(x)], \quad 0 \leqslant Fr(y) < 1 \qquad (9.26b)$$

We will approximate Fr(y) in Eq. (9.26b) by constructing a discrete table. The address or entry to the table during execution is determined by the K most significant bits to the right of the binary point of Co(x). The execution time on the CDC6400 is 16 μsec without interpolation and 26 μsec with interpolation (both times include the multiply to convert the logarithm to another base), in comparison to the system library function ALOG which requires 130 μsec. The rms error for $\log_e x$ is $0.20/N$ without interpolation and $0.020/N^2$ with interpolation. For N = 16 the rms errors are 0.012 and 7.8×10^{-5}, respectively, which amount to about .055 dB and .00035 dB for $10 \log_{10} x$.

Table 9.2/Errors for Discrete Computation of Square Root with Newton's Method				
		Relative Peak Error		
No. Iterations	N=8	N=16	N=32	N=64
0	3.1×10^{-2}	1.6×10^{-2}	7.8×10^{-3}	3.9×10^{-3}
1	5.0×10^{-4}	1.3×10^{-4}	3.1×10^{-5}	7.8×10^{-6}
2	1.3×10^{-7}	7.8×10^{-9}	5.0×10^{-10}	3.1×10^{-11}
3	7.8×10^{-15}	3.1×10^{-17}	1.3×10^{-19}	5.0×10^{-22}

A fast algorithm can be obtained by approximating $\log_2[Co(x)]$ by a linear function over the interval $1 \leqslant Co(x) < 2$. Since $\log_2 1 = 0$ and $\log_2 2 = 1$, the algorithm reduces to $Co(x) - 1$ so that

$$\tilde{y} = \log_2 x = Co(x) - 1 + Ex(x) \qquad (9.27)$$

where

$$In(\tilde{y}) = Ex(x) \qquad (9.27a)$$

$$Fr(\tilde{y}) = Co(x) - 1 \qquad (9.27b)$$

The term $[Co(x)-1]$ is just the part of $Co(x)$ to the right of the binary point. The execution time for this algorithm is 15 μsec on the CDC6400 and the rms error is 0.044 for $\log_e x$. The mean error for this algorithm is -0.036, which is comparable to the peak error. We can reduce the rms error by subtracting a bias of this amount in Eq. (9.27); however, we also increase the execution time slightly.

Power – $y = 2^x$

The base-2 power is again the logical choice for constructing algorithms. We can convert the base-2 power to any other base as $b^x = 2^{x/\log_b 2}$, which means that we rescale x before we execute the base-2 computation.

The power is the inverse of the logarithm. Therefore, let us express x as a fixed-point number and y as a floating-point number. Then we can write

$$y = 2^x = 2^{In(x)} \cdot 2^{Fr(x)} \qquad (9.28)$$

where

$$Ex(y) = In(x) \qquad (9.28a)$$

$$Co(y) = 2^{Fr(x)}, \qquad 1 \leqslant Co(y) < 2 \qquad (9.28b)$$

If $x < 0$, let us condition the fixed-point number so that $0 \leqslant Fr(x) < 1$. For this case we replace $In(x)$ by $In(x)-1$ and $Fr(x)$ by $1 + Fr(x)$. We will approximate $Co(y)$ by constructing a discrete table. The address or entry to the table during execution is given by the K most significant bits of $Fr(x)$. The execution time on the

CDC6400 is 16 μsec without interpolation and 26 μsec with interpolation (both times include the scaling of x), in comparison to the system library function EXP which requires 121 μsec. The relative rms error is 0.29/N without interpolation and $0.037/N^2$ with interpolation. For N = 16, the relative rms errors are about 2% and 0.01%, respectively.

We can again construct a fast algorithm by approximating $2^{Fr(x)}$ by a linear function over the interval $0 \leqslant Fr(x) < 1$. Since $2^0 = 1$ and $2^1 = 2$, the algorithm reduces to $Fr(x) + 1$ so that

$$Ex(\tilde{y}) = In(x) \qquad (9.29a)$$

$$Co(\tilde{y}) = Fr(x) + 1 \qquad (9.29b)$$

The execution time for this algorithm is 15 μsec on the CDC6400 and the relative rms error is about 6%. If we remove the bias (the mean error is .057), the relative rms error is reduced to less than 3% with a slight increase in execution time.

Sine-Cosine – $s = \sin 2\pi x$, $c = \cos 2\pi x$

Frequently in a simulation we must compute both the sine and cosine of a single argument. With the discrete approximation little additional effort is required to compute the second function.

First of all we must express x (which is the number of revolutions of a circle) as a fixed point word, again conditioned so that $0 \leqslant Fr(x) < 1$ regardless of the sign of x. In constructing the table for the discrete computation we will take advantage of $\cos 2\pi x = \sin 2\pi(.25-x)$ and the fact that all computations can be reflected to the first quadrant. Therefore we will construct a discrete table of sines $\{S_k\}$ by dividing the interval $0 \leqslant x \leqslant 0.25$ into N equal increments. During execution the two most significant bits of $Fr(x)$, which we designate as B1 and B2, determine the quadrant of the angle. The next K bits determine the address or entry to the table. If we designate k as the value of these K bits, the discrete sine and cosine are given by the following logic:

Quadrant	B1	B2	Sine	Cosine
1	0	0	$s = S_k$	$c = S_{N-1-k}$
2	0	1	$= S_{N-1-k}$	$= -S_k$
3	1	0	$= -S_k$	$= -S_{N-1-k}$
4	1	1	$= -S_{N-1-k}$	$= S_k$

where the index $N-1-k$ is just the complement of k. If interpolation is used, the remaining bits in $Fr(x)$ are converted to a fraction, h, to be used in Eq. (9.21). The execution time for the CDC6400 is 24 μsec without interpolation and 44 μsec with interpolation, in comparison to the system library functions SIN and COS which each require about 125 μsec. The rms errors are about $0.32/N$ without interpolation and $0.065/N^2$ with interpolation. If we use this procedure to generate the components of a phasor where the phase $2\pi x$ is given, the rms errors relative to the phasor amplitude will be somewhat larger as $0.45/N$ without interpolation and $0.091/N^2$ with interpolation.

In most video signal simulations we will not be able to tolerate a few-percent error in the computation of sine and cosine, since the sampling errors will appear as spurious noise in the signal spectrum. One specific case is the generation of Doppler-shifted signals and subsequent processing in the receiver. The general rule is to hold the rms error below the lowest sidelobe level of interest in a given situation. In practically all simulations we can tolerate an rms error of 10^{-4} since this will be 80 dB below the signal power. We can achieve this accuracy with $N = 4096$ without interpolation and $N = 32$ with interpolation. Since the former table size is probably excessive, we will be forced to implement interpolation to achieve the desired accuracy. If we relax the constraint to an rms error of 3×10^{-3} (-50 db), we can use a table size of $N = 128$ without interpolation.

The discrete algorithm can be made faster by extending the sine table $\{S_k\}$ over the interval $0 \leqslant x \leqslant 1.25$, which we then divide into $5N$ equal increments (to get the same accuracy as

before, this value of N is the same as the previous value). The address or entry to the sine table during execution is now determined by the K most significant bits of $Fr(x)$. Let k be this index. The address for the cosine is given by $k+N$ since $\cos 2\pi x = \sin 2\pi(x+.25)$.

Arctangent — $y = tan^{-1}x$

The arctangent is the logical choice for the inverse trigonometric function because of its near linear behavior over the interval $0 \leqslant x < 1$. For $x = 1$ we will simply set $tan^{-1}x = \pi/2$. For $x > 1$ we can set $tan^{-1}x = \pi/2 - tan^{-1}(1/x)$, and for $x < 0$ we can set $tan^{-1}x = -tan^{-1}(x)$. Over the unit interval we will approximate $tan^{-1}x$ by a discrete table. With x in a fixed-point format the address or entry to the table during execution is given by the K most significant bits of $Fr(x)$. The execution time on the CDC6400 is 19 μsec without interpolation and 29 μsec with interpolation, in comparison to the system library function ATAN which requires 135 μsec. The relative rms error is $0.25/N$ rad without interpolation and $0.022/N^2$ rad with interpolation. For $N = 16$, the rms errors are about 0.014 rad (.8°) and 9×10^{-5} rad (0.005°), respectively.

In all of the cases so far where no interpolation is used, certain reference computations such as $\sqrt{1}$, $\log 1$, e^0, $\sin 0$, and $tan^{-1}0$ result in the largest error. While we cannot reduce this error without increasing the table size, we can change the location at which the peak errors occur with only a slight increase in execution time. One way of accomplishing this result is to construct the discrete table values at the edge of each increment, instead of at the center. During execution the entry to the table is determined by rounding, instead of truncating. Then there will be no subsequent errors associated with the above reference computations.

References

[1]   Abramowitz, M., and I.A. Stegun; "Handbook of
       Mathematical Functions," *NBS Appl. Math. Ser.,
       Vol. 55*, June 1964, pp. 949-953.

[2]   Mitchell, R.L., and C.R. Stone; "Table-Lookup
       Methods for Generating Arbitrary Random Num-
       bers," *IEEE Trans. Computers*, scheduled for
       publication.

[3]   Aus, H.M., and G.A. Korn; "Table-Lookup/In-
       terpolation Function Generation for Fixed-Point
       Digital Computations," *IEEE Trans. Computers,
       Vol. C-18*, August 1969, pp. 745-749.

[4]   Ahrens, J.H., and U. Dieter; "Computer Methods
       for Sampling from the Exponential and Normal
       Distributions," *Comm. of ACM, Vol. 15, No. 10*,
       October 1972, pp. 873-882.

Chapter 10

Ground Based Radar Example

In the previous chapters we have discussed many techniques that can be applied to simulating radar signals, especially complex video ones. In this chapter we will show how these techniques can be applied to the specific problem of simulating a ground based search radar. As we emphasized in Chapter 2, the particular approach that best is suited to a radar simulation depends on many factors. Such a dependency becomes evident as we discuss the various options that are available. The final simulation procedures that emerge will be the result of several decisions made along the way. We have chosen an example that is complete in that all phases of signal generation and processing are discussed. This example is also intended to raise some new issues that have not yet been discussed. We should emphasize, however, that the resulting simulation is not meant to be applied universally; it is formulated about a specific objective and a specific radar system.

Table 10.1/System Parameters for Ground Based Search Radar		
Wavelength —	0.07 m	(λ)
Scan Rate —	60°/sec	($\dot{\theta}$)
Azimuth Half-power Beamwidth (two-way) —	3.2°	(θ_{3dB})
Elevation Beam —	fan	
Pulse Waveform —	13-bit binary Phase Code (Barker)	
Bit Length —	1.0 μsec	(T_B)
Pulse Length —	13.0 μsec	(T_p)
PRF (nominal) —	3.0 kHz	(f_r)
Doppler Processor —	Coherent FFT	
Number of Pulses —	16	(N)
Simulation Range Swath —	25-40 km	
Detection Threshold —	Constant False Alarm Rate based on average of ± 8 range cells	
Maximum Wind Velocity —	50 km/h	

10.1/Formulating the Problem

In this example we take an approach that will enable us to simulate the video signal with high accuracy, including all sidelobe effects. We begin by assuming that the simulation is to be performed for a ground based search radar at a specific site. Thus we need topographic maps or some other source of terrain elevation data. The first task in the simulation will be to convert this terrain data to a map of ground clutter RCS. We will need to do this only once since the clutter RCS map can be used as the input to future simulations. In addition we will simulate rain clutter and targets.

The Radar System

The radar system for this example will be based on a fan-beam antenna in elevation that scans continuously in azimuth over 360°. The basic system parameters are postulated in Table 10.1. One important parameter is the processing time required for detection. For this example it is the time to process N = 16 pulses, which is

$$T = \frac{N}{f_r} = 5.33 \text{ msec} \tag{10.1}$$

for the nominal PRF of 3 kHz. During this time the antenna will rotate through an angle of

$$\dot{\theta}T = 0.320° \tag{10.2}$$

Since this angle is much smaller than the one-way half-power beamwidth of 3.2°, the inequality in Eq. (2.15) is satisfied and we can assume that the antenna scan motion is frozen during the time it takes to process 16 pulses. Actually we could process as many as 40 pulses and still assume that the scan motion is frozen.

Sampling Rates

If the radar processes consecutive 16-pulse frames during the scan, the azimuth beams (each assumed to be frozen) will be separated by Eq. (10.2). However, in most search radars multiple PRF's are used to avoid blind speeds. Thus if the PRF changes from one 16-pulse frame to another, the azimuth beams will not be uniformly spaced around 360°. Other factors can affect the spacing of the beams; for example, the radar may perform tracking functions that are time-shared with search. For the purposes of this example let us assume that the search beams are separated by three times Eq. (10.2) or 0.960°. Thus the number of such beams over a 360° scan will be

$$N_{360} = \frac{360°}{3\theta_T} = 375 \qquad (10.3)$$

Since the angular separation of the beam is less than the half-power beamwidth, there will not be any significant errors introduced by our assumption that the beams are uniformly spaced, even if they are not.

In the range dimension we note that the bit length of 1 μsec corresponds to 150 m. In practically every receiver employing binary phase coded pulses the receiver response in range will be sampled at a rate corresponding to the bit length, or every 150 m for our example.

From the discussion above we have specified the sampling rates in range and angle for the simulation output. This step is a preliminary one that is required in order to implement the convolutions of the ground and rain clutter with the azimuth antenna pattern and the pulse modulation function. In some cases (e.g., an airborne radar simulation), it may be necessary to have a sampling rate internal to the simulation that is higher than that of the simulation output in order to reduce the sampling errors associated with the convolutions. However, for this example, we can have the same rate internally for the following reasons:

a) We will use the *reverse direction* to transform the terrain map into a ground clutter RCS map. In effect the RCS samples will be created on the proper range-azimuth grid and there will be no sampling errors associated with points off of the grid. This procedure is satisfactory as long as the grid is smaller than the structure of the terrain map. For this example the terrain can be specified on a grid no finer than 150 m (see Section 6.6).

b) The azimuth sample spacing is 30% of the two-way azimuth half-power beamwidth. This arrangement satisfies the requirement in Section 6.4 that there be at least 2 samples within the half-power width of the response (the two-way power pattern) in order to treat clutter in the sidelobe regions.

c) The range sample spacing corresponding to the bit length is adequate to treat clutter in the sidelobe regions for binary phase codes as we discussed in Section 6.4.

Thus we have established that the ground and rain clutter maps should be constructed on a polar coordinate grid, where the spacing is 0.960° in azimuth (375 samples around 360°) and 150 m in range. As we will soon see, the choice of 375 samples is a good one.

Number of Range Samples

The number of range samples on output over the 15 km interval of interest is 101, if we have a sample at the first and last ranges. We need additional range samples internal to the simulation because clutter just outside of the range interval of interest can affect detections inside through the pulse waveform sidelobes and the Constant False Alarm Rate (CFAR) processor. For the 13-bit phase code sampled at the bit length we need 12 additional range samples on each end, and the averaging of 8 range cells on each side of the cell under test to form the CFAR threshold means that we need 8 additional range samples on each end — bringing the total up to 141 range samples. The number of range samples at each stage of the simulation is illustrated in Fig. 10.1. We assume that the ground clutter is not range ambiguous, which limits us to a PRF of $f_r \leq 3750$ Hz for a maximum simulation of 40 km.

10.2/Clutter and Target RCS Descriptions

For this example the environment models will be a description of the RCS of ground clutter, rain clutter, and targets. The models of rain clutter and targets will be simple and expressed directly in radar coordinates; the ground clutter model will be a function of the topography of the radar site.

Site-Dependent Ground Clutter

To construct the ground clutter RCS map we begin with a topographic map that is digitized on, say, a 500 m rectangular grid. Since the maximum range in this simulation example is 40 km (if we neglect the multiple-time-around echoes), the topographic map need be no larger than a 160 x 160 matrix, or about 25,000 points. The points will represent the terrain elevation, although we could also describe the terrain type to be used for the calculation of RCS (elevation and terrain type could be packed into a single word). To transform elevation data in ground coordinates to polar, we proceed along one ray of constant azimuth using Eq. (6.51) to find the

---

Figure 10.1/Number of Range Samples at Various Stages in Simulation.

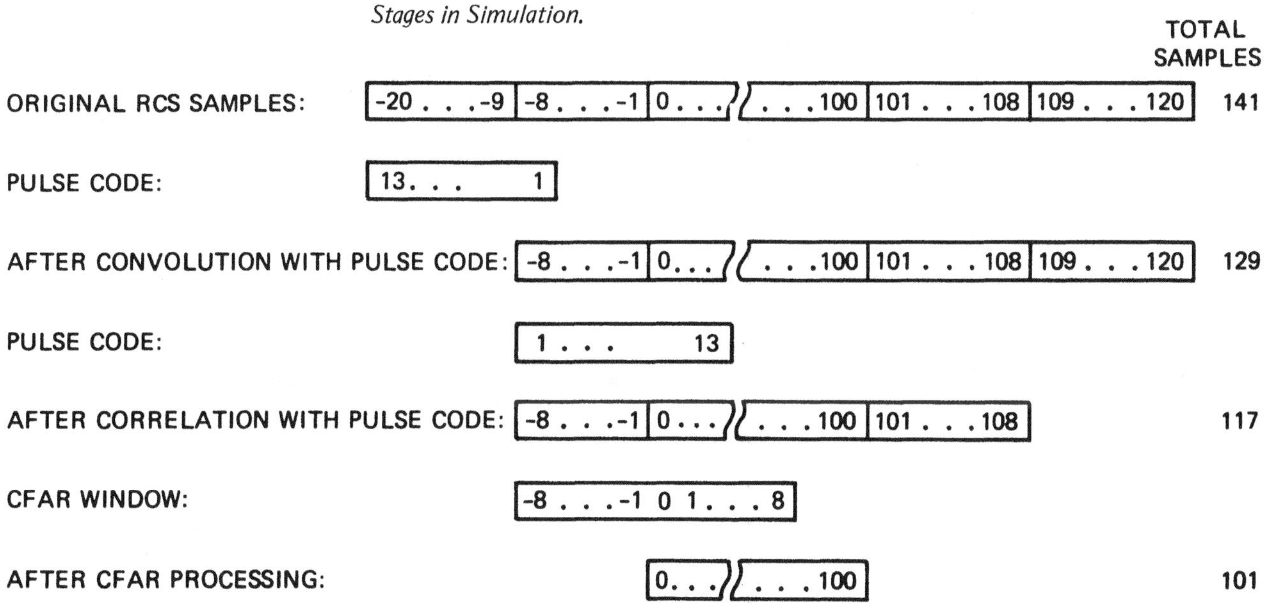

		TOTAL SAMPLES
ORIGINAL RCS SAMPLES:	-20 . . .-9 | -8 . . .-1 | 0 . . .⟩⟨ . . .100 | 101 . . .108 | 109 . . .120	141
PULSE CODE:	13 . . . 1	
AFTER CONVOLUTION WITH PULSE CODE:	-8 . . .-1 | 0 . . .⟩⟨ . . .100 | 101 . . .108 | 109 . . .120	129
PULSE CODE:	1 . . . 13	
AFTER CORRELATION WITH PULSE CODE:	-8 . . .-1 | 0 . . .⟩⟨ . . .100 | 101 . . .108	117
CFAR WINDOW:	-8 . . .-1 0 1 . . .8	
AFTER CFAR PROCESSING:	0 . . .⟩⟨ . . .100	101

(x, y) coordinate along the ray in uniform increments of $\Delta r = 150$ m. At each sample we can either go to the nearest terrain matrix element to find elevation, or we can interpolate on the surface (it is desirable to store the whole terrain matrix in core, but not necessary if we segment the computation). We must begin this process at some range closer than the minimum range of interest in order to include subsequent shadowing computations. The next step is to compute RCS at each sample from the elevation data (and terrain type if it is available) and to express it in the *model* of the ground clutter. For that part of the calculation, we can omit any reference to shadowing. The final step is to set the RCS of those samples that are shadowed to zero; this procedure is illustrated in Fig. 10.2. We will designate the resulting RCS samples as $\overline{\sigma}(\theta_\ell, \tau_m)$ at the respective discrete points in azimuth and delay (range). The samples will represent an ensemble average.

Antenna Pattern Weighting

It will be the objective of our example to simulate the video signal at the input to the radar processor. We could use Eq. (6.10) translated down to video to implement the various convolutions, but, to take advantage of the fixed sample spacing in azimuth for the search, we convolve the RCS first with the two-way azimuth power gain pattern. Since we have not yet defined the Doppler coordinate, let us write the ensemble average RCS weighted by the antenna as

$$\overline{|U(\theta, \tau_m)|^2} = \sum_\ell \overline{\sigma}(\theta_\ell, \tau_m)G^2(\theta_\ell - \theta) \qquad (10.4)$$

where θ takes on the discrete values of the azimuth beam pointing angles. This convolution is discrete and circular and it will be performed at each range sample. To implement this convolution it is best to do it indirectly by means of 375-point FFT's. The number 375 is rich in

prime factors (3 x 5 x 5 x 5) and the computation time to do the two necessary FFT's at each range ring is about 0.4 sec on the CDC6400.

Rain Clutter

If the rain clutter model were spatially homogeneous, we could treat the transformation to radar coordinates in about the same manner as we did for ground clutter. The basic differences are that shadowing is probably not significant and we must perform an integration in altitude over the elevation pattern of the antenna beam. For our purposes let us assume that the rain is homogeneous so that, with the assumption of a vertical fan beam, we can compute an equivalent RCS density for the rain clutter, as in Eq. (3.11). Moreover, since the clutter is homogeneous, we can also integrate the RCS over the two-way azimuth gain pattern. The resulting weighted RCS per sample, in the same units as Eq. (10.4), will be approximately

$$\overline{|U(\theta, \tau_m)|^2} = r_m\theta_{3dB}\Delta r\, G_o^2\, \sigma_o \qquad (10.5)$$

where r_m is the range to the mth sample, θ_{3dB} is the azimuth half-power beamwidth, Δr is the 150 m range sample spacing, G_o is the one-way gain on axis, and σ_o is the equivalent RCS density of the rain clutter as defined in Eq. (3.11). We note that the weighted RCS in Eq. (10.5) is independent of the beam direction θ (although the fluctuation spectrum will not be), and it is only weakly dependent on range for this example.

Data Formats

The format of the ground and rain clutter that we have just developed is especially suited for the simulation of a search radar, because the search pattern is uniform, or nearly so. The convolution of the RCS map with the antenna patterns prior to the generation of video signals is a big time saver since the convolutions have to be

Figure 10.2/Steps in Computing RCS from Terrain Elevation Data

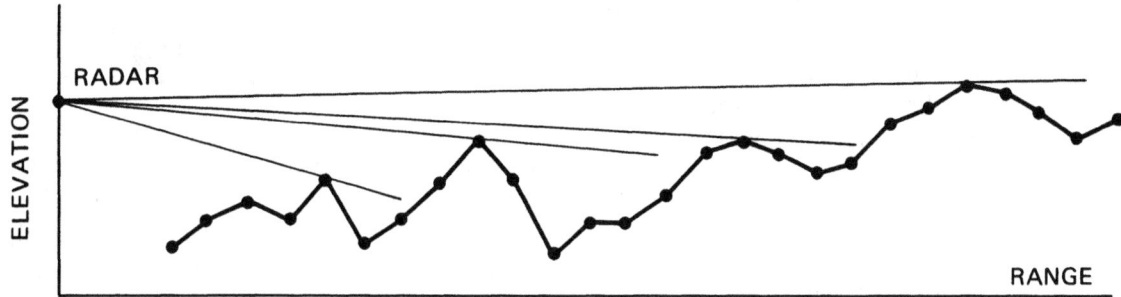

a. ELEVATION PROFILE

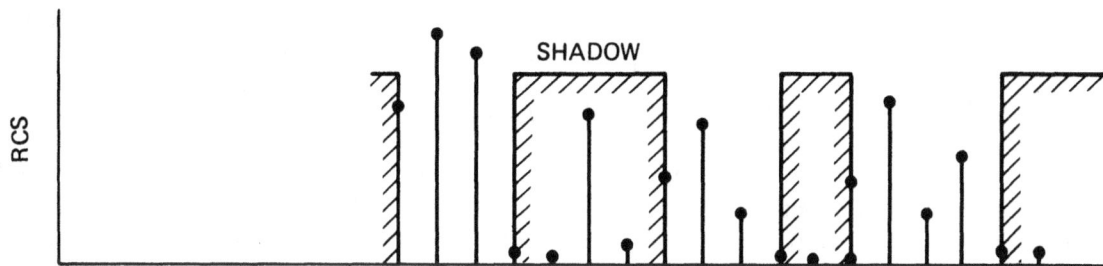

b. RCS PRIOR TO SHADOWING

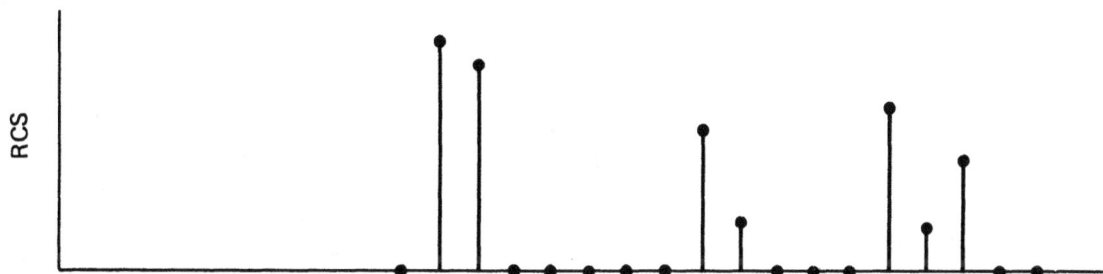

c. RCS AFTER SHADOWING

performed only once. However, the number of samples in the RCS matrix can be large. In this example the ground clutter matrix is 375 azimuth samples by 141 range samples, or almost 43,000 samples for the area of interest. Since we work with only one azimuth beam at a time, we can easily write the RCS matrix onto an auxiliary file (tape or disk), with one record for each azimuth beam. In this example the rain clutter is expressed as a mathematical function, which greatly simplifies the data handling problem.

Targets

In a typical simulation of a search radar a target will usually be expressed in statistical terms, such as a Swerling Case I target of a certain average RCS at some nominal range. However, with such a target description we would not need a site-dependent description of the ground clutter. In our example, we will assume that we are interested in specific target trajectories. Thus at any instant of time we can compute the azimuth angle of the target θ_T (relative to the radar), the range delay τ_T, and the ensemble average RCS at the particular target aspect, σ_T. Given the antenna gain pattern $G(\theta)$ we can write the weighted ensemble average power as

$$\overline{|U(\theta, \tau_T)|^2} = \sigma_T G^2(\theta_T - \theta) \qquad (10.6)$$

In most situations we will want to share the power between the adjacent range samples τ_m and τ_{m+1} (the samples separated by 150 m in this example) as

$$\overline{|U(\theta, \tau_m)|^2} = \sigma_T(1-\alpha)G^2(\theta_T - \theta)$$
$$\overline{|U(\theta, \tau_{m+1})|^2} = \sigma_T \alpha G^2(\theta_T - \theta) \qquad (10.7)$$

where α is the fractional distance that τ_T exceeds τ_m. For our example we will use Eq. (10.7).

Targets are of interest only if they appear in the mainlobe of the antenna. Therefore we neglect all targets appearing in antenna sidelobes, and Eq. (10.7) will be applied only within the mainlobe.

10.3/Generating the Video Signal

So far we have just performed transformations on the ensemble average RCS associated with ground clutter, rain clutter, and targets. We have not yet generated the coherent video signal, nor have we discussed the Doppler fluctuation of ground and rain clutter. Both are accomplished in the same step, as we will describe; also, we will add thermal noise to the received signal. Since the steps involved in generating and processing the video signal are repeated many times, we will give an estimate of the computation requirements for each simulation step.

Doppler Fluctuation

Let us assume that the ensemble average power $\overline{|U(\theta, \tau_m)|^2}$ given by Eq. (10.4) or Eq. (10.5) is actually distributed in Doppler as well as angle and delay. If we define $U(\theta, \tau_m, \nu_n)$ to be a random phasor assigned to the Doppler frequency ν_n in a Doppler cell of width $\Delta\nu$, then, if all such phasors are independent,

$$\sum_n \overline{|U(\theta, \tau_m, \nu_n)|^2} = \overline{|U(\theta, \tau_m)|^2} \qquad (10.8)$$

The first step in the simulation of the video signal is just the generation of the random phasors $U(\theta, \tau_m, \nu_n)$ at each range sample within each beam. The most efficient implementation for ground based radars is the 5-point approximation to the Gaussian-shaped spectrum derived in Section 8.4 and outlined in Fig. 8.6 — an approach that is generally valid as long as the fluctuation spectral width is not larger than the Doppler resolution, f_r/N, where N is the number of pulses coherently processed. A 50 km/h wind at a wavelength of $\lambda = 0.07$ m will have a maximum Doppler frequency of 400 Hz. The ground clutter spectral half-power width will be about 3% of this figure, or 12 Hz, as we discussed in Section 3.4, and we assume that the rain clutter spectral half-power width is 20% of the 400 Hz or 80 Hz. Both widths are much less than f_r/N = 188 Hz.

If we use the sample spacing of $\Delta\nu = 0.6\nu_{3dB}$ as given in Fig. 8.6, where ν_{3dB} is the half-power width of the assumed fluctuation spectrum ($\Delta\nu$ = 7.2 Hz for ground clutter and 48 Hz for rain clutter), we can generate the desired Doppler samples by utilizing the following procedure for each spectral component (ground or rain):

1/ Compute the amplitude $A = \sqrt{P}$, where $P = \overline{|U(\theta, \tau_m)|^2}$

2/ Generate 5 uncorrelated random phasors $\{\xi_n\}$, each of unit average power, with the desired amplitude distribution (see Chapter 9)

3/ Form the products
$$U(\theta, \tau_m, \nu_1) = .1020 \; A\xi_1$$
$$U(\theta, \tau_m, \nu_2) = .4558 \; A\xi_2$$
$$U(\theta, \tau_m, \nu_3) = .7508 \; A\xi_3$$
$$U(\theta, \tau_m, \nu_4) = .4558 \; A\xi_4$$
$$U(\theta, \tau_m, \nu_5) = .1020 \; A\xi_5$$

When we substitute these random phasors into Eq. (10.8), we can easily verify the equality. We repeat this procedure at each range sample, τ_m, for the ranges of interest; for this example, there are 141 range samples in total. This process must be implemented separately for ground and rain clutter because each spectrum generally is centered at a different Doppler frequency. If a target is present it will be represented by a single spectral line located at the Doppler frequency of the target. If ground or rain clutter and one target are all present in a range ring, there will be 11 Doppler samples in total, as we illustrate in Fig. 10.3. Note that the samples are positioned where the spectral energy is located. Thus they are nonuniformly spaced.

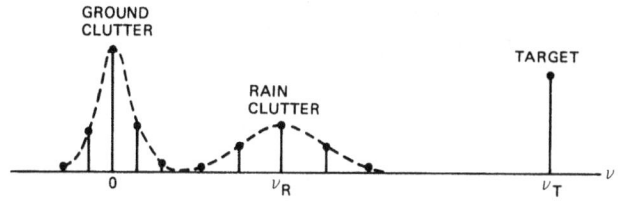

To estimate the computation time required to generate the 5 spectral components we assume that the table-lookup methods of Chapter 9 are used. Each phasor ξ_n will require about 28 to 35 μsec on the CDC6400, depending on the amplitude distribution; there is one square-root that requires about 13 μsec; and there will be 8 or 10 multiplies of 5.5 μsec each, depending on how the procedure is implemented. Thus an upper limit is about 110 μsec on the CDC6400 for the 5 spectral components, including certain overhead computations. For ground and rain clutter the 141 x 10 matrix of samples in each azimuth beam can be generated in about 160 msec.

We are now in a position to write the received signal and the receiver response for this example. Using the notation in Fig. 6.2 for coherent pulse trains we can write

$$\psi_R(\theta, t, \ell T_r) =$$
$$e^{j2\pi f_c(t+\ell T_r)}\sum_m \sum_n \alpha_m U(\theta, \tau_m, \nu_n)\mu_{PT}(t-\tau_m)e^{j2\pi\nu_n\ell T_r}$$

$$(10.9)$$

and

$$Z(\theta, \tau, \nu) = \sum_m \sum_n \alpha_m U(\theta, \tau_m, \nu_n)\chi_P(\tau-\tau_m)D(\nu-\nu_n)$$

$$(10.10)$$

where we have assumed that all pulses are weighted uniformly on transmit (a_ℓ = 1 in Fig. 6.2). The summations are over a more limited region than in that figure because the range interval for simulation purposes is much less than the range to the first ambiguity, and the Doppler domain can be described by as few as 5 samples. If we translate Eq. (10.9) down to video, assuming that the receiver is completely coherent, we can write the received video signal as

$$\mu_R(\theta, t, \ell T_r) = \sum_m \sum_n \alpha_m U(\theta, \tau_m, \nu_n)\mu_{PT}(t-\tau_m)e^{j2\pi\nu_n\ell T_r}$$

$$(10.11)$$

It is this expression that we implement for our example. For reference, θ is the particular azimuth of the beam for which we are simulating the received video signal, t is the time measured from the transmission of each pulse, and ℓ is the pulse number. In our example t will be sampled at every 1.0 μsec corresponding to the 150 m sample spacing and ℓ will take on any 16 consecutive integers, which we will assign $\ell = 1$, ..., 16. The parameter α_m is a range-dependent scale factor given by Eq. (6.8). However, as we discussed in Section 6.4, it is desirable to make the sampling errors unbiased. In this case the samples in the range dimension contribute the greatest error. For Barker codes sampled at the bit length we must multiply the RCS of all samples by 2/3 to remove the sampling bias. Thus we can define

$$\alpha_m^2 = \frac{2}{3} \frac{\lambda^2}{(4\pi)^3 r_m^4} \qquad (10.12)$$

or

$$\alpha_m = .0183 \lambda/r_m^2 \qquad (10.13)$$

To save time we do the scaling by α_m prior to the generation of the random Doppler phasors. The product $\alpha_m U(\theta, \tau_m, \nu_n)$ is then treated as a single phasor.

The quantity $\mu_{PT}(t)$ in Eq. (10.11) is the transmitted pulse modulation function. With the scaling of Eq. (10.12), $\mu_{PT}(t)$ must be scaled at the actual transmitter power. If we designate P_T as the peak transmitter power, then

$$\max|\mu_{PT}(t)|^2 = P_T \qquad (10.14)$$

We could have included the factor of P_T in Eq. (10.12), in which case $\mu_{PT}(t)$ would then be

scaled to unit peak amplitude. We could also include the system losses, either in the scaling of $\mu_{PT}(t)$ or in Eq. (10.12); however, we use the convention of Eq. (5.94) where the system losses are added to the receiver noise power. Note that if the propagation losses are nonhomogeneous, they must be included in the definition of the product $\alpha_m U(\theta, \tau_m, \nu_n)$ in Eq. (10.11).

Discrete Convolution and DFT

Of the two summations in Eq. (10.11) the one in range is a discrete convolution and the one in Doppler is a Discrete Fourier Transform (DFT). Which one should be done first is the next question to resolve. Let us refer to Fig. 10.1 where we have listed the number of range samples at various stages of processing. We observe that each stage of processing reduces the number of samples. In Doppler we can have as few as 5 samples or as many as 11, depending on what combination of clutter and targets is present in a range ring. The DFT converts Doppler to the sequence of 16 pulses. Thus this step actually *increases* the number of samples, so that if the DFT were performed first we would have to implement 16 separate convolutions in range, instead of the 11 or fewer that are required if we do the convolutions before the DFT's. In general the DFT's should be done first if the number of pulses, N, is less than the number of clutter samples. Otherwise the range convolutions should be implemented initially.

Clutter samples tend to exist everywhere on the 2-D range-Doppler grid (some of the ground clutter may be shadowed, although we usually cannot take advantage of this fact if the convolutions in range are performed first). On the other hand, if a target is present in the azimuth beam, it will be represented by only two samples (when we use Eq. (10.7) to share the power between adjacent range samples). If we add a row to the 2-D grid at the Doppler of the target, we would only increase the matrix to 11 Doppler samples if both ground and rain clutter are also present. Even though practically all of the samples in that row will be zero, it simplifies the procedure just to add the row to the matrix when a target is present, instead of implementing a special computation.

For the 141 x 10 matrix of ground and rain clutter phasors $U(\theta, \tau_m, \nu_n)$, we implement the discrete convolution in range first. At each Doppler sample, represented by the index n, we compute

$$\eta_n(t_k) = \sum_{m=-20}^{120} \alpha_m U(\theta, \tau_m, \nu_n)\mu_{PT}(t_k - \tau_m). \qquad (10.15)$$

$$t_k = t_o + k\Delta t \qquad t_o = \tau_o = 166.67 \ \mu sec \ (25 \ km)$$

$$\tau_m = \tau_o + m\Delta\tau \qquad \Delta t = \Delta\tau = 1.0 \ \mu sec \ (150 \ m)$$

$$k = -8, \ldots, 120$$

Then we do the DFT's at each of the remaining 129 range (time) samples represented by the index k as

$$\mu_R(\theta, t_k, \ell T_r) = \sum_{n=1}^{10} \eta_n(t_k)e^{j2\pi\nu_n\ell T_r}, \qquad \ell = 1, \ldots, 16$$
$$k = -8, \ldots, 120$$
$$(10.16)$$

where

Ground Clutter	Rain Clutter
n $= 1, \ldots, 5$	n $= 6, \ldots, 10$
$\nu_n = \nu_3 + (n-3)\Delta\nu$	$\nu_n = \nu_8 + (n-8)\Delta\nu$
$\nu_3 = 0$	$\nu_8 = -(2V_W/\lambda)\cos(\theta - \theta_W)$
$\Delta\nu = 0.6\nu_{3dB}$	$\Delta\nu = 0.6\nu_{3dB}$
	θ_W = wind direction
	V_W = wind velocity

While the summation in Eq. (10.15) is written over 141 samples, the length of $\mu_{PT}(t)$ is only 13 samples. Thus for each of the 129 output samples, there will be only 13 basic computations if we implement the convolution directly. At each Doppler sample row the computation time will be about 129 x 13 x 40 μsec = 68 msec on the CDC6400, assuming that the product $\alpha_n U(\theta, \tau_n, \nu_n)$ is formed first and $\mu_{PT}(t)$ is complex. That time can be cut in half for our example because $\mu_{PT}(t)$ is a binary phase code and can be treated as a scalar. Moreover, we can eliminate the multiplications by ±1 in the phase

code if we scale α_m^2 by the peak transmitter power; such action, though, will not save a substantial amount of time. For our example we assume that the computation time associated with each Doppler sample row is about 35 msec, or 350 msec for the 10 Doppler sample rows. If we implement the convolutions in Eq. (10.15) indirectly via 2 FFT's of length 256 (probably the optimum length), the computation time for each Doppler sample row is 2 x 256 x 8 x 20 μsec = 82 msec, or 820 msec for the 10 Doppler sample rows. The indirect method is slower than the direct method, but not by a substantial amount. If the pulse modulation function were complex and slightly longer, the opposite conclusion would be reached.

Since Eq. (10.16) has a nonuniform sample spacing over the PRF interval, we are forced to implement the DFT in a brute-force manner. The computation time for each range sample column, where there are 10 input Doppler samples and 16 output pulse samples, is 16 x 10 x 40 μsec = 6.4 msec on the CDC6400. For the 129 range sample columns the time is 830 msec. If we tried to do the discrete Fourier transform via an FFT, we would need one long enough to resolve the spectrum of the ground clutter into at least two parts. An FFT of length 512 and a PRF of 3000 Hz would have a Doppler sample spacing of about 6 Hz, or 50% of the half-power width of the ground clutter spectrum. For each range sample column the computation time would be 512 x 9 x 20 μsec — over 14 times longer than the brute-force implementation. In addition we would have to generate more than 5 Doppler samples for the rain clutter because its spectrum is much broader than the ground clutter spectrum. The advantage of using a nonuniform sample spacing in Doppler for ground based radars is illustrated quite clearly. If we use the best methods available, we can generate the received signal for one beam of ground and rain clutter in less than 1.35 sec on the CDC6400.

For each target present in the beam we add one more sample to the index n in Doppler. The

Doppler frequency of a target is $\nu_T = -(2V_T/\lambda) \cdot \cos(\theta-\phi_T)$ where V_T and ϕ_T are the target velocity and heading, respectively. For each Doppler sample that we add to the original 10, we increase the computation time in direct proportion. For one target the increase is only 10%, which is tolerable; even up to 6 targets the increase in computation time is acceptable when we consider that most of the azimuth beams do not have any targets at all. As we exceed 16 Doppler samples it begins to be more attractive to implement the DFT's before the range convolutions. To reduce the total computation time we could segment the computation into two parts — the first would be done on the 10 Doppler samples of clutter, convolving in range and then implementing the DFT's in Doppler as discussed above; the second part would be performed on the remaining Doppler samples of the targets. If there are K targets, we would construct a 141 x K matrix of the samples $U(\theta, \tau_m, \nu_k)$ where most would be zero. Then, by testing for zeros in the range convolutions (a direct implementation) and the DFT's before the multiplications are performed, we can reduce the overall computation time to about 10 msec per target. This computation time would be negligible compared to the 1.35 sec for ground and rain clutter.

Thermal Noise

Thermal noise can be added almost anywhere prior to the first nonlinearity. The two most convenient places to add it are − 1) to the Doppler samples $U(\theta, \tau_m, \nu_n)$, *provided the Doppler samples are uniformly spaced throughout the PRF interval*; and 2) to the received pulse samples $\mu_R(\theta, t_k, \ell T_r)$ in Eq. (10.16), or at any point prior to the receiver processing but after the DFT from Doppler to the pulse sequence. Since the Doppler samples are nonuniformly spaced in our example, we will implement the second procedure.
Considering that we are adding the thermal noise prior to pulse compression, we have to calculate

what the noise power is at that point. First of all, let us assume that we implement the pulse compression as a sampled version of Eq. (5.76) as

$$Z_\ell(\tau) = \Delta t \sum_k \mu_R(\theta, t_k, \ell T_r)\mu_{PF}^*(t_k-\tau) \qquad (10.17)$$

where $\{\mu_{PF}(t)\}$ is a sequence of pulse waveform samples to which the receiver is matched. Then, for a thermal noise sequence $\{n(t_k)\}$ substituted for $\mu_R(\theta, t_k, \ell T_r)$, and after taking the ensemble average, we obtain the power after pulse compression as

$$P_N = \overline{|n(t_k)|^2}\, (\Delta t)^2 \sum_k |\mu_{PF}(t_k)|^2 \qquad (10.18)$$

Now let us define

$$p_n = \overline{|n(t_k)|^2} \qquad (10.19)$$

as the ensemble average power in a noise sample, and

$$E_{PF} = \Delta t \sum_k |\mu_{PF}(t_k)|^2 \qquad (10.20)$$

as the energy in the waveform to which the received pulses are matched. Then

$$P_N = p_n \Delta t\, E_{PF} \qquad (10.21)$$

However, from Eq. (5.91), $P_N = N_o E_{PF}$, where N_o is the noise power density. Therefore, we can write the ensemble average power in a noise sample associated with each sample $\mu_R(\theta, t_k, \ell T_r)$ as

$$p_n = N_o/\Delta t \qquad (10.22)$$

where Δt is the spacing of the samples t_k. Since we can generate a Gaussian random number in 14 μsec on the CDC6400 (see Section 9.2), the time required to add the thermal noise samples to the 129 x 16 matrix of pulse samples, including the scaling by Eq. (10.18), will be less than 100 msec.

We have completed the generation of the received video signal, including thermal noise, for one azimuth beam. The next step is to simulate the processing in the receiver.

10.4/Receiver Processing

The signal at the input to the receiver for one azimuth beam consists of a sequence of N = 16 pulse phasors for each of the 129 range samples. If the processing were linear in the receiver, we could implement the steps prior to detection in any order; however, with linear processing we should have simulated the receiver response in Eq. (10.10) directly, instead of first generating the received video signal. Therefore we assume that there is some nonlinearity at IF or video that precedes the steps of Doppler filtering and pulse compression (e.g., the pulse compression is implemented digitally). Thus the procedure is to simulate the various processing steps in the receiver in the order in which they occur. For our example we assume that these steps are 1) Stability and Time Control (STC) scaling, 2) clipping and A/D conversion, 3) pulse compression, 4) Doppler filtering, 5) square-law detection, and 6) CFAR processing.

STC Scaling

There is usually some range or STC scaling step implemented in the receiver initially, to reduce the dynamic range of target signals. Since the received power from a target is inversely proportional to r^4 where r is range, a typical scaling function is r^4. If we designate r_o as some reference range, then all received pulse samples would be scaled as

$$\left(\frac{r_k}{r_o}\right)^2 \mu_R(\theta, t_k, \ell T_r)$$

according to the r^4-power law, where $r_k = ct_k/2$.

Clipping and A/D Conversion

Clipping at IF is implemented on the phasor magnitude of each received pulse sample. In a

slight deviation from Eq. (5.85) let us define C to be the power clipping level and $v_R + jv_I$ to be one pulse phasor sample. The clipping procedure is given by

$$\rho^2 = (v_R^2 + v_I^2)/C$$

if $\rho^2 \leqslant 1$, no clipping will be performed;

if $\rho^2 > 1$, then take a square-root and set

$$v_R = v_R/\rho$$

$$v_I = v_I/\rho$$

Even if a square-root is done on every pulse sample, the time requried to implement this procedure will be less than 60 μsec on the CDC6400 if the fastest table look-up algorithm for the square-root in Section 9.2 is used. For the 129 x 16 matrix of pulse samples, the total time will be less than 120 msec; in most situations, we will need only about half of that time — 60 msec.

If the A/D converter has K bits, the voltage quantization step q in Fig. 5.5 should be related to the clipping power level as

$$q^{K-1} = \sqrt{C} \qquad\qquad (10.23)$$

so that the highest level in the A/D converter will not be exceeded on input. In most cases, q is set according to the noise (and/or clutter power). The A/D conversion is implemented separately on each phasor component, v_R and v_I, according to an integer-arithmetic operation. For the A/D converter in Fig. 5.5 the operation is truncation downward on the quantities (v_R/q) and (v_I/q). About 25 μsec per phasor sample is required on the CDC6400 to do this integer arithmetic, or about 50 msec for the 129 x 16 matrix.

Pulse Compression

The remaining processing steps prior to detection, but after the nonlinearity, can be done in any order. We implement pulse compression first because there will be 12 fewer range samples to be processed subsequently. For our example the pulse compression is identical to the discrete convolution implemented in Section 10.3, except that, instead of a convolution, we have a correlation. As before, the best implementation for our example is the direct method. For each pulse sample row there will be 117 output range samples, as we show in Fig. 10.1, each requiring 13 basic operations. The computation time for each pulse sample row is 117 x 13 x 20 μsec = 31 msec on the CDC6400; for all 16 pulse sample rows, the time is 500 msec.

Doppler Filtering

The Doppler processor for our example is a 16-pulse FFT. To reduce the Doppler sidelobes the pulses will be weighted prior to the FFT. For each range sample column requiring one FFT the computation time will be about 1.6 msec on the CDC6400, including the weighting; for the 117 range sample column, the time will be about 190 msec. The ouptut format will be 16 Doppler filter samples (filter outputs) at each of the 117 range samples.

Square-Law Detection

For our example the detection step will consist only of computing the power (sum square of the phasor components) in each Doppler filter sample. The comparison with the threshold will be implemented in the next step. The computation time to compute the power in all 117 x 16 samples is about 30 msec on the CDC6400. If a linear detector were employed we could use a (max + 1/2 min)-type of operation to estimate the phasor amplitude, or we could use a fast square-root algorithm from Section 9.2. The former calculation would require about 20 msec for the 117 x 16 matrix; the latter, about 45 msec.

CFAR Processing

In our example the detection threshold for any sample under test is formed by first averaging the detected power in 8 range samples on either

side of the sample under test, or 16 total samples. Let us designate P_{kn} as the detected power in the kth range sample and n^{th} Doppler filter sample. The detection threshold for the sample P_{kn} is then

$$T_{kn} = \frac{C}{16}\left[\sum_{m=k-8}^{k+8} P_{mn} - P_{kn}\right] \quad (10.24)$$

where C is a threshold multiple or CFAR constant, which is a function of the desired false alarm rate. The detection test is then just the comparison $P_{kn} : T_{kn}$. This test will be implemented over 101 range gates and 16 Doppler filters for our example, although the outputs of the filters near dc will probably be ignored in further processing because they contain clutter. The computation time for this CFAR processing algorithm is less than 40 msec on the CDC6400.

In Table 10.2 we have summarized the steps required to process the video signal for one azimuth beam. All computation times are referenced to the CDC6400 computer, and we have assumed that the ground clutter RCS map has already been computed and written to a file. Clearly, the operations of DFT and convolution dominate the computation time; therefore, it is especially important that efficient algorithms be made available for those two operations — assuming that the fastest random number generators are used in Steps 3 and 6. Without the table-look-up algorithms described in Section 9.2 those two steps could take as much as twenty times longer to execute, and thus be the most time-consuming — each requiring about as much time as the whole simulation now takes.

For one scan consisting of 375 beams we need approximately 1000 sec of computation time — about 160 times more than in real time.* In that length of time we will have simulated 375 x 101 x 16 $\simeq$ 606,000 detected samples at the

*We have simulated only 30% of the ranges and 33% of the azimuth beams during one scan. If everything were simulated the computation would be another factor of 10 slower than real time.

Table 10.2/Summary of Steps in Generating and Processing Video Signal for One Azimuth Beam.

Step	Range	Matrix Size Doppler	Pulses	Computation Time (msec)	
1. Read Ground Clutter RCS Map	141			20	
2. Compute Rain Clutter RCS	141			10	
3. Generate Random Phasors	141	10		160	~1.5 sec
4. Range Convolution (10 times)	129	10		350	
5. DFT to Generate Pulses (129 times)	129		16	830	
6. Add Thermal Noise	129		16	100	
7. STC Scaling	129		16	40	
8. Clipping & A/D Conversion	129		16	110	
9. Pulse Compression (16 times)	117		16	500	~1.0 sec
10. Doppler Filtering (117 times)	117	16		190	
11. Square-Law Detection	117	16		30	
12. CFAR Processing	101	16		40	
Total Computation Time per Beam					~2.5 sec

output of 101 range gates and 16 Doppler filters. With only ground clutter present we have to blank (ignore) the Doppler filter that is centered at dc; in this case, we have 375 x 101 x 15 ≃ 568,000 false alarm opportunities. With rain clutter we have to blank at least one more Doppler filter output; if we know which filter contains the clutter, we can blank just that one (the Doppler spread of the rain clutter is about 80 Hz for our example, compared to the Doppler filter width $f_r/N = 188$ Hz). Sometimes, though, the rain clutter will straddle two filters, in which case both must be blanked. In such an instance, unless the dc-filter also contains the rain clutter, three filters would have to be blanked; there would be 375 x 101 x 13 ≃ 492,000 false alarm opportunities in one scan.

10.5/Recording Detections

In the preceding discussion we showed that there is a considerable amount of computation involved in generating video signals. Instead of performing a threshold test for a predetermined value of C in Eq. (10.24) or for a fixed threshold if no CFAR is implemented, it is expedient to accumulate detection statistics as a function of the threshold setting. This operation is best accomplished by computing a histogram of the detected samples (first normalized by the reference power in the case of CFAR processing). By integrating the histogram to obtain the sample distribution function we are able to determine the detection performance for any threshold setting. We illustrate this result in Fig. 10.4 where we have recorded the simulation statistics for three different targets and the false alarms. The abscissa is the threshold setting (in arbitrary units) and the ordinate is the percentage or fraction of samples that have exceeded the specified threshold setting. By plotting the results on a Gaussian probability scale, the high and low ends of the graph are expanded to permit a more accurate interpretation.

To show how results plotted as in Fig. 10.4 can be utilized, suppose the system spec is for a false alarm probability of 10^{-4}. If we read across the graph at that value to the curve for false alarms, then down to the abscissa, we have the threshold setting that corresponds to the desired false alarm probability, namely 3.5 units. Now we can read the probability of detecting the three targets at that threshold setting as 0.65 for Target 1, 0.80 for Target 2, and 0.95 for Target 3. The various targets could represent different radar cross sections, ranges, or target headings.

We can easily determine the precision with which we can simulate a given probability of a threshold crossing. If we assume that the simulated samples are independent, the number of threshold crossings out of a total sample size of N will be binomially distributed. Thus if we designate the probability of a threshold crossing as p, the mean number of threshold crossings is Np and the variance is Np(1−p). The *fractional* mean and variance are p and p(1−p)/N, respectively. Thus the standard deviation of the estimate of p is

$$\sigma_p = \sqrt{p(1-p)/N} \qquad (10.25)$$

As an example — if N = 100 and p = 0.50, then σ_p = 0.05, or 10% of p. On the other hand if we made p = 0.10 for the same number of samples, then σ_p = .03, or 30% of p. To get σ_p down to 10% of p, we would need N = 900 samples. To show what happens as we decrease p still further, let us write the ratio

$$\frac{\sigma_p}{p} = \sqrt{\frac{1-p}{Np}} \qquad (10.26)$$

Now if we neglect the 1−p since p is small, we can write

$$Np = 1/(\sigma_p/p)^2 \qquad (10.27)$$

Thus we see that, in order to maintain a fixed ratio of σ_p/p, we must keep the product Np constant.

A small value of p in the above equations corresponds to estimating the false alarm probability. Although we can establish many different criteria for specifying N, a relatively simple one is to keep σ_p/p fixed. If we choose to make σ_p/p = 30%, then we can select N according to

$$N = 10/P_{fa} \qquad (10.28)$$

where P_{fa} is the desired false alarm probability. To simulate a false alarm probability of $P_{fa} = 10^{-4}$, we should have N = 10^5 samples. For detection probabilities it would make more sense to fix the ratio $\sigma_p/(1-p)$. If we maintain the same relative precision, then we should choose

$$N = 10/(1-P_D) \qquad (10.29)$$

where P_D is the desired detection probability. This action assumes that P_D is close to unity; if it is not, we should derive N from Eq. (10.25). Thus to simulate a detection probability of P_D = 0.90, we should choose N = 100. Note the penalty for specifying too great a precision. If the relative precision were 10% instead of 30%, we would need 10 times as many samples for both cases above. While the 1000 samples required to simulate the detection probability would not involve much cost, the 10^6 samples required to simulate the false alarm probability would. For this chapter's example we would need 20 scans to obtain enough samples — a process that would take over 5 hours of computing on a CDC6400.

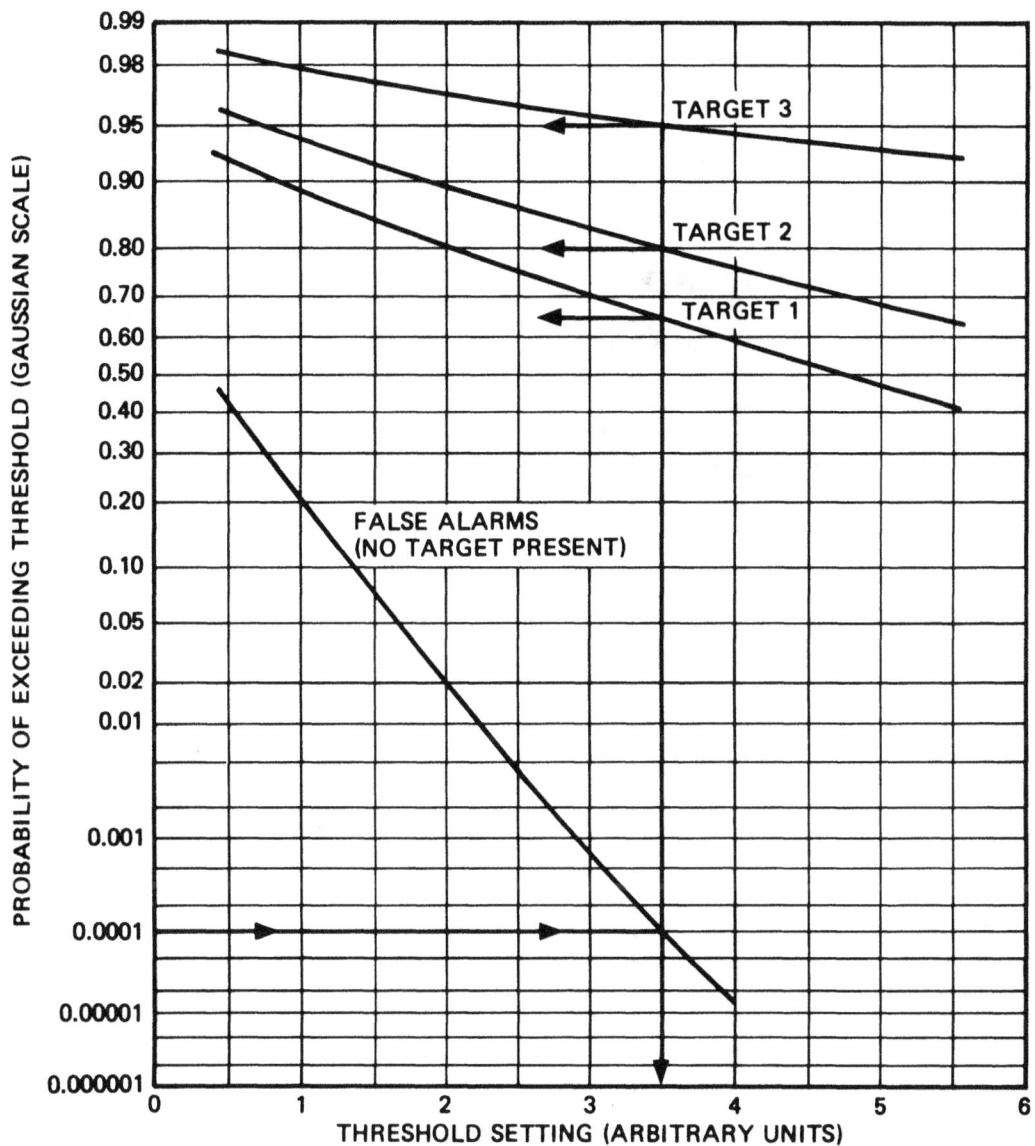

Figure 10.4/Sample Probabilities of Detection and False Alarm as Functions of the Threshold Setting

Chapter 11

Airborne Radar Example

To illustrate some of the problems associated with simulating airborne radars we have chosen the simulation of ground clutter to an airborne search radar. We will concentrate on how ground clutter gets transformed into radar coordinates, specifically for the case where the clutter is non-homogeneous; We assume that the processing in the receiver is linear up to the detector. The key system parameters for this example are given in Table 11.1. For simplicity we assume that the flight path is horizontal and that the antenna pattern is in an azimuth-elevation coordinate system.

Table 11.1/System Parameters for Airborne Search Radar

Wavelength —	0.03 m	(λ)
Antenna Azimuth Beamwidth (One-Way) —	2°	(θ_{3dB})
Waveform —	Coherent Pulse Train	
Pulse Length —	20 μsec	(T_p)
Pulse Bandwidth —	5.0 MHz	(B)
Spacing of Range Samples —	30 m	
PRF —	10 kHz	(f_r)
Doppler Processor —	Coherent FFT	
Number of Pulses —	128	(N)
Aircraft Altitude —	5 km	(h)
Aircraft Velocity —	240 m/sec	(V)

11.1/Clutter Model

We assume that the ground clutter is nonhomogeneous; however, as can be the case for most airborne search radars, the model will be statistical. The clutter backscatter coefficient σ_o (RCS per unit area on the ground) is constant throughout a square of about 1 km x 1 km on the ground (the size is chosen arbitrarily), but will vary from one square to another, independently, according to a log-normal distribution.

To account for the variation of σ_o with the grazing angle γ we define the spatial median of the distribution to be composed of two components as

$$\sigma_o(\gamma)_{median} = \sigma_{od} \sin\gamma + \sigma_{os}e^{-(90°-\gamma)^2/\phi_o^2} \qquad (11.1)$$

where the first term on the right is a diffuse component and the second is a specular component that accounts for the well-known altitude return in airborne radars. We show an example of Eq. (11.1) in Fig. 11.1. The constants σ_{od} and σ_{os} are deterministic quantities ($\sigma_{od} \ll \sigma_{os}$), as is ϕ_o, the peak-to-(1/e) width of the specular component. In general ϕ_o is about 5° depending

on the type of terrain. To complete the definition of the log-normal statistic, we must specify σ_L, the standard deviation of the log-variate. For our purposes let σ_L = 5 dB, or 5/4.343 = 1.15 in natural units. Since the ratio of mean to median for the log-normal distribution is $e^{-\sigma_L^2/2}$, we can rewrite Eq. (9.10) as

$$\ell = e^{g\sigma_L} \qquad (11.2)$$

to be the transformation from a Gaussian random sample g of zero mean and unit variance to a log-normal sample ℓ of unit *median* value. Within any 1 km x 1 km square on the ground we generate one log-normal sample ℓ and scale it by Eq. (11.1) according to the grazing angle of that square.

The ground clutter cells are defined in polar coordinates, as in Fig. 6.12. For our example, though, we do not have to generate the complete clutter map at one time; we need only one ring of clutter at any given moment. Therefore, we can choose the angular width of a clutter cell, θ, to vary inversely with range so that the product $r\theta$ is the constant of 1 km (or approximately 1 km, in order to make $2\pi/\theta$ an integer). The ground range width of the ring will also be

1 km. To distinguish this clutter cell from one we will define later, we will designate this one as the *large clutter cell*.

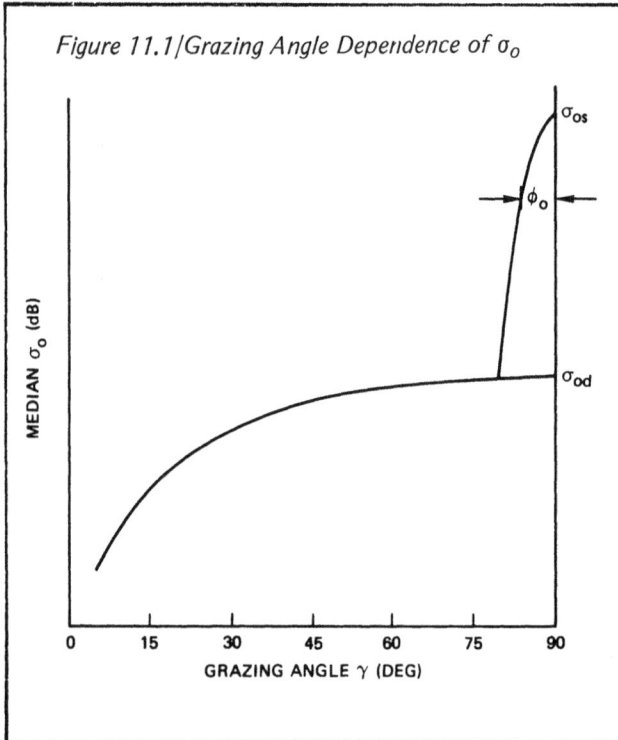

Figure 11.1/Grazing Angle Dependence of σ_0

11.2/Gridded Cell Structure

For the purpose of mapping clutter to radar coordinates we need to establish two different grids (a total of three if we count the large clutter cell derived above). Onto one grid in ground coordinates, we construct a matrix of discrete scatterers; that grid should be fine enough to make individual scatterers basically unresolvable. The second grid is in the radar coordinates of range and Doppler; the scatterers on the first grid are mapped onto this one.

Range Geometry

For our example we simulate all possible ranges at the output of the receiver. Because of the coherent pulse-train waveform, the maximum output range is the ambiguous interval defined by

$$r_{AMB} = \frac{c}{2f_r} = 15.0 \text{ km} \tag{11.3}$$

for the PRF of 10 kHz. Since the receiver must be off while each pulse is being transmitted, there is a minimum range on output which corresponds to

$$r_{MIN} = c(T_P + T_s)/2 \tag{11.4}$$

where T_P is the pulse length and T_s is the time required to perform certain switching functions. However, clutter at shorter ranges, including ambiguous ones, can affect the receiver output through the range processing sidelobes. Thus we have to simulate every sample within r_{AMB}, unless the desired swath of the simulation output is less than $r_{AMB} - 2(cT_P/2)$. For our example we assume that $T_s = 5$ μsec and $T_P = 20$ μsec, which means that $r_{MIN} = 3.75$ km.

The range sample spacing on output is 30 m, which is identical to the range resolution in terms of the bandwidth as

$$\Delta r_{RES} = \frac{c}{2B} = 30 \text{ m} \tag{11.5}$$

Since we must have at least two samples within this distance to simulate clutter in the waveform sidelobe regions, and since the ratio of the output and internal sample spacings should be an integer, we define the range sample spacing internal to the simulation as

$$\Delta r = \Delta_{RES}/2 = 15 \text{ m} \tag{11.6}$$

The number of samples within the ambiguous range interval is given by

$$N_\tau = r_{AMB}/\Delta r = 1000 \tag{11.7}$$

We should select the samples so that one of them falls at r_{MIN}. In our example $r_{MIN}/\Delta r = 250.0$, so we can adopt the convention that the mth range sample is given by $r_m = (m-1)\Delta r$ for $m = 0, \ldots,$

$N_\tau - 1$. On output we are interested only in the samples beginning with m = 250, and only in every other sample (since the sample spacing on output is half of that on input). There will be 375 output range samples in our example.

The range to the radar horizon is given by Eq. (6.49) as

$$r_{HOR} = \sqrt{2r_e h} = 291 \text{ km} \qquad (11.8)$$

where $r_e = 8500$ km is the radius of the earth for propagation purposes and h is the platform altitude of 5 km. While the ground clutter extends well into the 20th range ambiguous zone, the principal contributions, at least for clutter in the antenna sidelobes, is from the first few zones. Since the mainbeam of the antenna may not intersect the ground except at long ranges, it is possible to have ground clutter in the mainbeam from any or all of the 20 range ambiguous zones. To reduce the computational burden it is expedient to divide the ground into two regions, as we show in Fig. 11.2. In the antenna sidelobe region we simulate the clutter from only the first three range ambiguous zones, plus an additional amount to account for the fact that, for r < h, there is no clutter in the first zone. Thus the maximum unambiguous range of the sidelobe clutter is

$$r_{MAX} = 3r_{AMB} + h = 48.0 \text{ km} \qquad (11.9)$$

The sidelobe clutter that is omitted by this operation would have contributed less than 1% to the total clutter RCS in any one range cell. We can use Eq. (11.9) as long as h < r_{AMB}; if it is not, then we should set $r_{MAX} = 3h$.

The angular width of the mainlobe sector should be approximately the null-to-null width of the azimuth beam, or slightly larger if the close-in sidelobes of the antenna are high. For our example this angular width is about 4°. Instead of

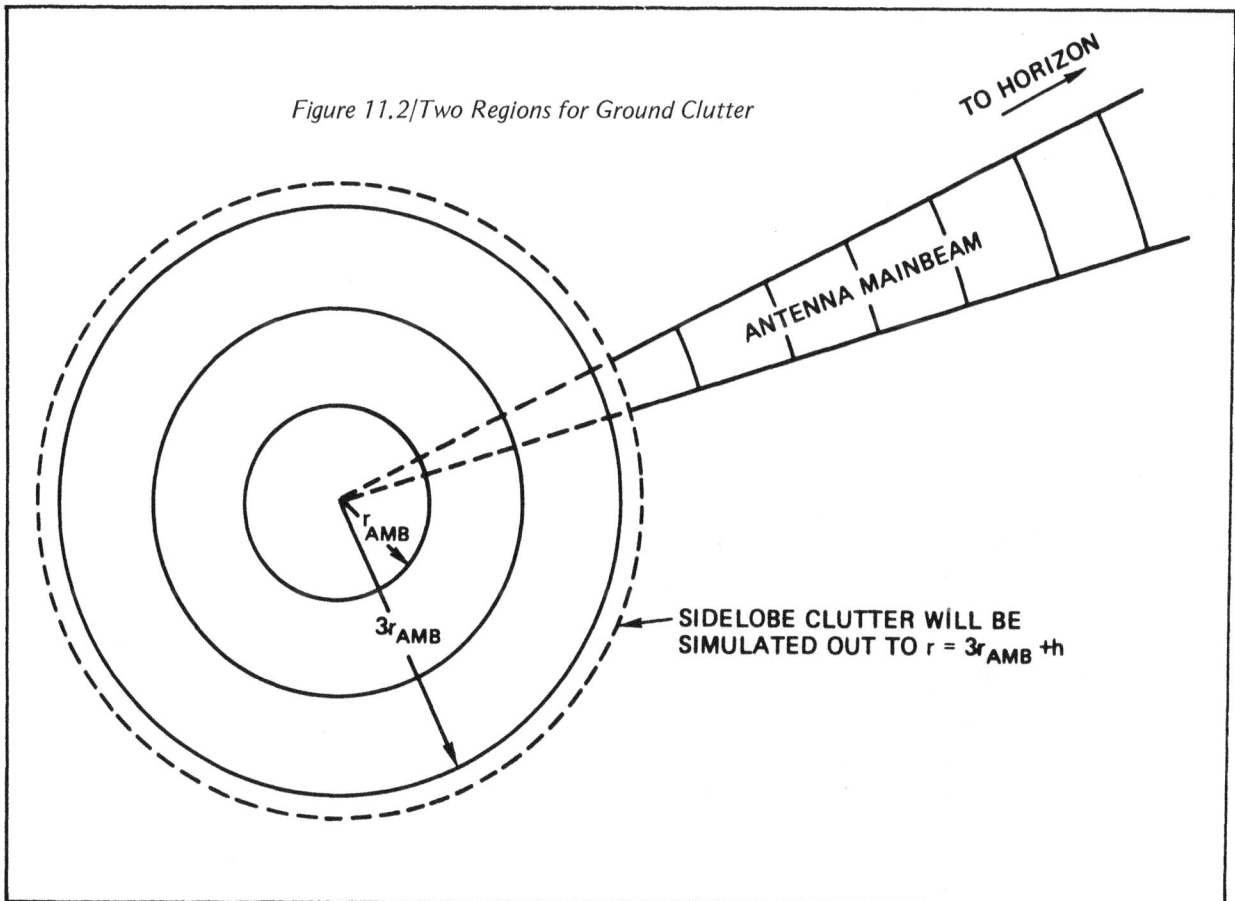

Figure 11.2/Two Regions for Ground Clutter

ANTENNA MAINBEAM

TO HORIZON

r_{AMB}

$3r_{AMB}$

SIDELOBE CLUTTER WILL BE SIMULATED OUT TO r = $3r_{AMB}$ +h

simulating the ground clutter from all range ambiguous zones within the mainlobe sector, we can make a calculation to see which zones are illuminated by the elevation beam. However, this process will not save much computation time since most of that time nonetheless is spent in the sidelobe region.

Geometry of the Small Clutter Cells

We map ground clutter in the forward direction for airborne radars; that is, we establish a clutter cell on the ground and assign a scatterer to it, the RCS of which describes the total for that cell. We then determine the range and Doppler for that scatterer and map the RCS *into* radar space. As we discussed in Section 6.8, this procedure allows us to keep the area of the clutter cell constant. If we were to do the mapping in the reverse direction, the calculation of the cell area would be nontrivial. The cell we are discussing now is designated as the *small clutter cell*.

First we have to establish the area of the cell. For the moment let us confine the range to be larger than the radar altitude by at least some small amount — say, approximately 5 times the range sample spacing. At the range sample r_m in Fig. 11.3 the ground range is given by

$$\rho_m = \sqrt{r_m^2 - h^2} \;\; , r_m > h \qquad (11.10)$$

where h is the radar altitude. The width of the range ring on the ground is

$$\Delta\rho_m = r_m \Delta r / \rho_m \;\; , r_m > h \qquad (11.11)$$

If we use the spherical-earth approximations from Section 6.8, the grazing angle $(\gamma_m = \beta)$ is given by Eq. (6.48), which we rewrite here as

$$\sin\gamma_m = h/r_m - r_m/2r_e \qquad (11.12)$$

This function is needed for Eq. (11.1). The elevation angle measured at the radar is

$$\epsilon_m = -\sin^{-1}(h/r_m + r_m/2r_e) \qquad (11.13)$$

which we need for the antenna pattern weighting. These parameters are constant for the whole range ring.

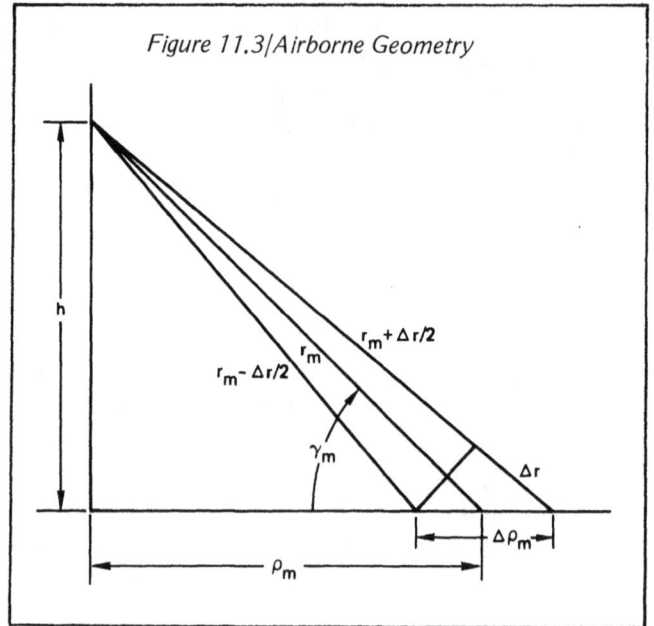

Figure 11.3/Airborne Geometry

The next step is to subdivide the range ring into a number of equal size angular sectors. To determine the angular width of each we have to establish what the Doppler sampling rate is internal to the simulation. If we process N pulses coherently via an FFT of length N, the Doppler filters are separated by $\Delta\nu = f_r/N$, where f_r is the PRF. However, to integrate clutter in the Doppler sidelobes we need at least two samples in that interval. Thus we define the Doppler sample spacing internal to the simulation to be

$$\Delta\nu = \frac{f_r}{2N} \qquad (11.14)$$

which, for our example of f_r = 10 kHz and N = 128, is $\Delta\nu$ = 39.0625 Hz. The number of samples within the PRF interval is

$$N_\nu = \frac{f_r}{\Delta\nu} = 2N \qquad (11.15)$$

which is 256 samples for our example. Given $\Delta\nu$ in Eq. (11.14) we can solve for the maximum angular width of the small clutter cell in Eq. (6.46) as

$$\Delta\theta \leqslant \frac{\lambda}{2V} \frac{f_r}{2N} \qquad (11.16)$$

In terms of the number of these cells around a range ring we can write

$$N_\theta = \frac{2\pi}{\Delta\theta} \geqslant 4\pi N \frac{2V}{\lambda f_r} \qquad (11.17)$$

The quantity $2V/\lambda f_r$ is the ratio of the maximum Doppler and the PRF. For our example where $V = 240$ m/sec, $\lambda = .03$ m, and $f_r = 10$ kHz,

$$\frac{2V}{\lambda f_r} = 1.6 \qquad (11.18)$$

Since this ratio exceeds unity, the PRF interval will be filled with clutter at all but the shorter ranges. Now, if we substitute Eq. (11.18) and $N = 128$ into Eq. (11.17), we obtain for our example

$$N_\theta \geqslant 2573 \qquad (11.19)$$

To make it a round number, we choose $N_\theta = 2600$. We now define

$$\Delta\theta = 2\pi/N_\theta \qquad (11.20)$$

The area of each small cell is given by

$$\Delta A_m = \rho_m \Delta\rho_m \Delta\theta = r_m \Delta r \Delta\theta \qquad (11.21)$$

which is a constant for all cells in the range ring. We define the azimuth angle of the center of the kth cell as

$$\theta_k = (k-1)\Delta\theta , \quad k = 1, \ldots, N_\theta \qquad (11.22)$$

RCS of Small Clutter Cell

The area of the small clutter cell is given by Eq. (11.21). Therefore, the RCS of the cell defined by the coordinate (ρ_m, θ_k) is

$$\sigma_{mk} = \sigma_o \Delta A_m \qquad (11.23)$$

where σ_o is the backscatter coefficient of the larger clutter cell that contains the polar coordinate (ρ_m, θ_k). This RCS will be assigned to a point scatterer located at (ρ_m, θ_k).

Doppler of Small Clutter Cell

Let us begin with Eq. (6.42) with $\beta = \gamma_m$ so that $\cos\beta = \cos\gamma = \rho_m/r_m$ and $\sin\beta = \sin\gamma_m = h/r_m$.

Then we can write the Doppler frequency ($\nu = 2\dot{r}/\lambda$) of the coordinate (ρ_m, θ_m) as

$$\nu_{mk} = \frac{2V}{\lambda}\left[\frac{\rho_m}{r_m}\cos\delta\cos\theta_k + \frac{h}{r_m}\sin\delta\right] \qquad (11.24)$$

where δ is the platform dive angle measured below the horizontal. Note that the products $(2V/\lambda)(\rho_m/r_m)\cos\delta$ and $(2V/\lambda)(h/r_m)\sin\delta$ are constants for a whole range ring. Moreover, we can precompute $\cos\theta_k$, so that Eq. (11.24) can be reduced to one addition and one multiplication.

11.3/Mapping Procedure

The general procedure is to generate large clutter cells within a ring that contains the ground range ρ_m, as we show in Fig. 11.4. As we increase k around the ring, σ_o changes about every 1 km according to the fluctuated value in the larger cell. A single computation involving integer arithmetic allows us to find σ_o to be used in Eq. (11.23). As we increase ρ, we can retain the same sequence of values of σ_o in the larger cells until ρ exceeds the outer boundary; then we can generate a new sequence of σ_o values for the next ring of larger cells. Thus we need only one sequence at any one time, except when dealing with the multiplicity of range ambiguities.

There are multiple ambiguous ranges that correspond the the range sample r_m in the first zone as

$$R_{m\ell} = r_m + \ell r_{AMB} , \quad \ell = 0, \ldots, L \qquad (11.25)$$

where L can be as large as 19 for the mainlobe clutter in our example. While mapping scatterers into the mth range sample, it is expedient to do so for all range ambiguities as well. The reason becomes evident when we discuss data formats.

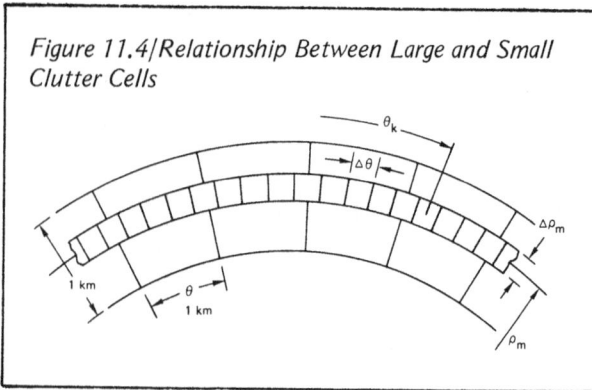

Figure 11.4/Relationship Between Large and Small Clutter Cells

At each ambiguous range sample r_m, we establish an array of samples of length $N_\nu + 1 = 2N+1$ into which we map the RCS of the Doppler samples in Eq. (11.24). For our example, $N_\nu + 1 = 257$. Let us designate this array as the *PRF-array*. The samples in it will be spaced by $\Delta\nu$ in Eq. (11.14), with the Doppler of the nth sample in the array given by

$$\nu_n = n\Delta\nu, \qquad n = 0, \ldots, N_\nu \qquad (11.26)$$

The PRF-array corresponds to one PRF interval, with one sample at the beginning of the interval and one at the end. The ambiguous Doppler frequency of ν_{mk} is

$$\nu'_{mk} = \nu_{mk}(\text{mod } f_r) \qquad (11.27)$$

The indices for the two array samples that surround ν'_{mk} are n and n+1 where

$$n = [\nu'_{mk}/\Delta\nu] \qquad (11.28)$$

The brackets [ ] designate "the largest integer less than". If we combine Eqs. (11.15), (11.24), (11.27), and (11.28), we can write

$$n = \left[N_\nu \frac{2V}{\lambda f_r}\left(\frac{\rho_m}{r_m}\cos\delta\cos\theta_k + \frac{h}{r_m}\sin\delta\right)\right](\text{mod } N_\nu)(11.29)$$

The quantities $N_\nu(2V/\lambda f_r)(\rho_m/r_m)\cos\delta$ and $N_\nu(2V/\lambda f_r)(h/r_m)\sin\delta$ are constants for one ring. For level flight we can write

$$n = \left[N_\nu \frac{2V}{\lambda f_r}\frac{\rho_m}{r_m}\cos\theta_k\right](\text{mod } N_\nu) \qquad (11.30)$$

The next step is to weight the RCS, σ_{mk}, by the parameters in the radar range equation, including the power gain $G(\theta_k, \epsilon_m)$ in the direction of the scatterer, as

$$\overline{|V_{mk}|^2} = \frac{G^2(\theta_k, \epsilon_m)\lambda^2}{(4\pi)^3 r_m^4}\sigma_{mk} \qquad (11.31)$$

The final step is to share the power in Eq. (11.31) between the two adjacent samples in the PRF-array, n and n+1. When we finish mapping all clutter cells in a ring and all subsequent range ambiguous rings, we must add the last sample ($n = N_\nu$) to the first ($n = 0$) since they both have the same ambiguous Doppler frequency. The resultant array is now of length N_ν.

Data Formats

For our example we have $N_\tau = 1000$ ambiguous range samples and $N_\nu = 256$ Doppler samples in the PRF-array. This total is over 250,000 samples, which exceeds the core size in practically all computers. Even if all samples could fit in core, it would be advantageous to complete the mapping of one range ring of clutter and its ambiguities into one PRF-array and then write the array to auxiliary storage. The same array can be reused for the next range ring. The disadvantage in this approach is that, sooner or later, we have to process the data as a function of range. In order to do so, we have to first transpose the matrix in auxiliary storage.

Computation Time

For sidelobe clutter in our example there are $2600 \times 1000 = 2.6 \times 10^6$ small clutter cells per ambiguous range zone, or 7.8×10^6 cells for the 3 zones. On the other hand there are only $30 \times 1000 = 3.0 \times 10^4$ small clutter cells per ambiguous range zone in the 4° wide mainlobe region, or about 6×10^5 cells for the remaining 19 zones. Thus there are over 15 times as many cells in the sidelobe as in the mainlobe region. In all there are about 8.4×10^6 small clutter cells that have to be transformed into radar space.

The computation time is usually dominated by the antenna pattern calculation in Eq. (11.31), especially if a coordinate transformation to a system other than (θ, ϵ) has to be performed. Considering a basic multiply time of about 5 μsec on the CDC6400, it is estimated that at least 50 μsec will be required to transform each sample into radar space — the computations involved are Eqs. (11.30) and (11.31), and the reverse interpolation used to share power between the two samples in the PRF array; these considerations are based on a very simple calculation for the antenna pattern. Even with optimum procedures the time could easily be 100 μsec per sample if interpolation were required in the antenna pattern calculation. Since there are 8.4×10^6 samples in our example, the computation time will range from 420 to 840 sec for one transmission (N = 128 pulses); thus we see the necessity to use optimum procedures to map the ground clutter into radar space.

11.4/Altitude Return

Directly beneath the radar on the ground is a narrow circular region that reflects the radar signal much more strongly than does the area slightly outside of this region. In our example this specular reflection is described by a Gaussian shaped function in Eq. (11.1). If we define the nadir angle as $\phi = 90° - \gamma$, then the backscatter coefficient for this specular term is

$$\sigma_{os}(\phi) = \sigma_{os}e^{-(\phi/\phi_0)^2} \qquad (11.32)$$

According to the geometry of Fig. 11.5 the area on the ground bounded by the two rings at cone angles ϕ and $\phi + \Delta\phi$ is

$$\Delta A = 2\pi(h\tan\phi)(h\Delta\phi/\cos^2\phi) \qquad (11.33)$$

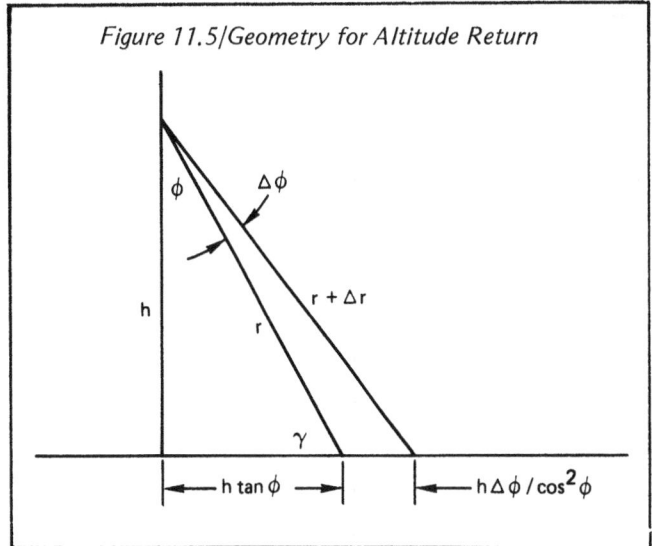

Figure 11.5/Geometry for Altitude Return

But since $h = r\cos\phi$ and $\Delta r = r\tan\phi\,\Delta\phi$, we can write

$$\Delta A = 2\pi r \Delta r \qquad (11.34)$$

where, for the moment, we do not associate Δr with the range sample spacing. The total RCS appearing within the range increment Δr is thus

$$\sigma_s(r)\Delta r = \sigma_{os}(\phi)\Delta A \Big|_{\phi = \cos^{-1}(h/r)}$$

$$= 2\pi r \Delta r \sigma_{os}(\phi) \Big|_{\phi = \cos^{-1}(h/r)}$$

For small values of ϕ

$$\phi = \sqrt{2(r-h)/h}\,, \qquad r \geqslant h \qquad (11.35)$$

so that from Eq. (11.32) we can write

$$\sigma_s(r)\Delta r = 2\pi r \Delta r \sigma_{os}e^{-2(r-h)/h\phi_0^2}\,, \quad r \geqslant h$$

$$= 0 \qquad\qquad\qquad r < h \qquad (11.36)$$

Thus the RCS as a function of range is just a biased exponential function, beginning at r = h. If we replace Δr by dr and integrate over all ranges, the total RCS for the altitude return is

$$\sigma_s = \int_h^\infty \sigma_s(r)dr = \pi(h\phi_0)^2\sigma_{os} \qquad (11.37)$$

The mean range for the altitude is where $\sigma_s(r)$ is down $1/e$ from its peak, or

$$\bar{r} = h(1+\phi_0^2/2) \tag{11.38}$$

For our example we assume that $\phi_0 = 0.1$ rad. Thus for h = 5000 m, we find that $\sigma_s = 7.8 \times 10^5 \sigma_{os} m^2$ and $\bar{r} = 5025$ m. Usually σ_{os} is close to unity, which means that the total RCS of the altitude return can approach 60 dBm².

If $\bar{r}-h \ll \Delta r$, where Δr is now the range sample spacing internal to the simulation, we can approximate the altitude return as being concentrated at $\bar{r}$. In this case the RCS is given by Eq. (11.37). However, if $\bar{r}-h \geqslant \Delta r$, we should sample the altitude return as in Eq. (11.36), with Δr being the range sample spacing internal to the simulation. Assuming that the range samples are at the center of the cells, some special computation should be made for the first cell that contains the altitude return (i.e., the mth range sample where $|r_m-h| \leqslant \Delta r/2$ because that cell will not be completely filled with clutter. The altitude return decays rapidly with range and we can terminate the computation when it is no longer the dominant term. Usually this situation occurs when $r \geqslant h + 4(\bar{r}-h)$, at which time the return is about 20 dB down from its peak. In our example $\bar{r}-h = 25$ m and $\Delta r = 15$ m, so we should compute the altitude return over about 6 or 7 range samples.

The Doppler spectrum of the ground clutter is also straightforward to estimate. Let us begin with Eq. (6.42) with $\beta = 90°-\phi$, and write the Doppler frequency $(\nu = -2\dot{r}/\lambda)$ as

$$\nu = \frac{2V}{\lambda} \left(\cos\theta \sin\phi \cos\delta + \cos\phi \sin\delta \right) \tag{11.39}$$

where δ is the dive angle of the flight vector relative to the horizontal. For the altitude return from a narrow ring at range r we can set $\sin\phi$ equal to ϕ in Eq. (11.35) and $\cos\phi = 1$. The maximum and minimum Doppler frequencies occur at $\theta = 0$ and $\theta = \pi$, respectively; the central or mean frequency, at $\theta = \pi/2$ or $3\pi/2$. Thus we

can write the central frequency as

$$\nu_o = \frac{2V}{\lambda} \sin\delta \tag{11.40}$$

and the peak deviation as

$$\nu_p = \frac{2V}{\lambda} \cos\delta \sqrt{2(r-h)/h} \tag{11.41}$$

For level flight $\nu_o = 0$ and

$$\nu_p = \frac{2V}{\lambda} \sqrt{2(r-h)/h} \tag{11.42}$$

If we assume that the clutter is uniformly distributed around the ring, the shape of the Doppler spectrum is

$$S(\nu) = \frac{1}{\pi \sqrt{\nu_p^2-(\nu-\nu_o)^2}} \tag{11.43}$$

This function is normalized so that the integral is unity as

$$\int_{\nu_o-\nu_p}^{\nu_o+\nu_p} S(\nu)d\nu = 1 \tag{11.44}$$

For our example where the platform is at an altitude of 5 km, the diameter of the circle on the ground within the cone angle of $\phi_0 = 0.1$ rad is 1 km, which is about the size of the larger clutter cell. Thus for all practical purposes we can assume that the clutter is homogeneous, at least for the first few range samples. In generating the altitude return we would simply make σ_{os} in Eq. (11.36) a constant, possibly selected initially from a random process. For most simulation applications we will lose little generality if we extend this assumption on spatial homogeneity over all range samples that contain the altitude return. The exceptions, of course, are those radars that operate specifically on the altitude return, such as a radar altimeter.

11.5/Point Clutter

Scattering from man-made objects tends to be much more intense than the scattering from terrain. In general it is not possible to resolve individual scatterers on any one object; therefore, we treat the clutter as a point-scattering phenomenon. To simulate point clutter we can generate a ground coordinate (ρ, θ) and an RCS for each point according to some distribution function, or we can have a deterministic model for the object. In many cases the points are concentrated in groups corresponding to the cultural features in or near towns and industrial complexes; in rural areas or the wilderness the points tend to occur at random.

As long as there are not too many points, say less than about 1000, we should generate the coordinates and RCS of all points first and sort them in ambiguous-range order. Then, as we are doing the mapping of the distributed clutter for one range sample, we can examine the range of the nearest point clutter sample to see if it is within the range ring. If it is, we should share its power (RCS) among the four nearest samples in the 2-D range-Doppler matrix as in Fig. 6.3. We can accomplish this result in two passes, one for each range sample.

11.6/Processing the Signal

So far we have discussed how to create a matrix of samples in the radar coordinates of range and Doppler that describes the distribution of RCS (or ensemble average signal power) associated with ground clutter. If targets are present, we could add the target RCS to the proper clutter samples, just as we did with point-clutter. But if we do so, we have to assume that the amplitude statistics associated with the targets are the same as those for the clutter. If any of the components have different distribution functions, we must treat them separately until we have generated the complex video signals.

There are just two major steps to processing the range-Doppler power matrix for a linear receiver. The first is the creation of the complex video signal; the second is the 2-dimensional convolution with the waveform ambiguity function, which is listed in Fig. 6.2 for the coherent pulse train and slightly rewritten here as

$$Z(\tau, \nu) = \sum_{m=0}^{N_\tau - 1} \sum_{n=0}^{N_\nu - 1} V_{mn} \chi_P(\tau - \tau_m) D(\nu - \nu_n) \qquad (11.45)$$

In processing the signal we assume that the power matrix is in auxiliary storage, each record corresponding to one range sample containing the PRF-array. Since the processing in the receiver is assumed to be linear, we rearrange the processing steps so that the number of accesses to auxiliary storage is minimized. First we process in the Doppler dimension.

Generating the Video Signal

For the mth ambiguous range sample, the Doppler samples in the power matrix are given by a sequence of scalars $|V_{mn}|^2$, $n = 0, \ldots, N_\nu - 1$. First we create N_ν independent random phasors V_{mn}, where each has the specified average power. While the phasor amplitude could be distributed according to any distribution function, it is usually satisfactory to assume that the signal fluctuations within a given cell are Rayleigh amplitude distributed, in contrast to the way in which the average RCS varies from cell to cell. With the Rayleigh assumption the components of the random phasor are independent Gaussian samples.

For each random phasor we must compute a square-root, generate two random numbers, and do two multiplies. If we use the fast algorithms discussed in Chapter 9, the computation associated with each phasor is about 60 μsec on a CDC6400 computer. For the N_ν = 256 Doppler samples in each range ring the computation time is about 15 msec, and for all N_τ = 1000 range samples it is 15 sec.

Convolution with Doppler Fine Structure

The Doppler fine structure $D(\nu)$ repeats with a period f_r. Therefore, the convolution in Doppler is *circular*, represented by

$$Y(\tau_m, \nu) = \sum_{n=0}^{N_\nu - 1} V_{mn} D(\nu - \nu_n) \qquad (11.46)$$

It is best implemented indirectly via two fast Fourier transforms. The first transform is of length N_ν; the second is of length $N = N_\nu/2$, where N is the number of pulses processed in the receiver. We can illustrate this fact if we first write the pulse sequence

$$
\begin{aligned}
y_m(\ell T_r) &= \sum_{n=0}^{N_\nu - 1} V_{mn} e^{j 2\pi \ell n T_r \Delta \nu} \\
&= \sum_{n=0}^{N_\nu - 1} V_{mn} e^{j 2\pi \ell n / N_r} \qquad (11.47)
\end{aligned}
$$

where we have assumed that all transmitted pulses have the same amplitude. While the sequence in ℓ is of length N_ν, we now select any N consecutive samples and implement the Doppler filtering as

$$Y(\tau_m, \nu_k) = \sum_{\ell=0}^{N-1} \omega_\ell y_m(\ell T_r) e^{-j 2\pi k \ell / N}, \quad k=0, \ldots, N-1$$

$$, m=0, \ldots, N_\tau - 1$$

$$(11.48)$$

where the new Doppler samples ν_k are separated by f_r/N. The sequence $\{\omega_\ell\}$ is the sequence of pulse weights used to reduce the Doppler sidelobes.

If we implement our example on a CDC6400 computer, the computation time for $N = 128$ is

about (256 x 8 + 128 x 7) x 20 μsec for the FFT's, and about 3 msec for the multiplication by real weights. Since this computation must be repeated at all N_τ = 1000 range samples, the total computation time is slightly over 100 sec.

Transposing the Matrix

After each Doppler column on the range-Doppler matrix is processed, we again write the result onto an auxiliary file. After all columns are processed, we must implement a convolution in range according to Eq. (11.45). However, the format of the auxiliary storage is not compatible with this operation. We must first transpose the matrix.

The transpose step is basically very simple if we use multiple passes through the matrix. Suppose we have a working area of 30,000 words available in core (which means that we can process 15 rows at one time for our example, each containing 1000 complex examples — or 30 rows where the samples are real). For each pass through the first matrix we transfer the samples in each column that correspond to the 15 rows in the partial matrix. When the partial matrix is filled after one pass, it is written onto a second auxiliary file, one row per record. The first file is rewound and the above steps are repeated for the next 15 rows. Since we have N = 128 rows in our example, we require nine passes to complete the transposition (four passes would be sufficient for a working area of 1000 x 64 = 64,000 words). This general procedure is illustrated in Fig. 11.6 for a simple example.

To minimize the number of passes, the working area in core should be made as large as is practical. The computation time is determined mostly by the read/write access time for the particular peripheral devices. Since the utilization of auxiliary storage is highly dependent on the particular installation, it is difficult to estimate a cost associated with its use. For the purposes of this discussion we assume that the matrix transpose is equivalent to about 80 sec of computation on a CDC6400 computer for a problem of our size. On any system, however, it will probably be cheaper to implement the matrix transpose than to process the 1000 pairs of FFT's involved in the previous step.

Convolution with Pulse Autocorrelation Function

The convolution in Eq. (11.45) with the pulse autocorrelation function $\chi_P(\tau)$ is also *circular*.

Figure 11.6/Steps in Transposing a Matrix on an Auxiliary File (illustrated for a 7 x 9 matrix with 27 samples of working area available in core).

FILE 1; 9 RECORDS, 7 SAMPLES IN EACH RECORD
FILE 2; 7 RECORDS, 9 SAMPLES IN EACH RECORD
WORKING AREA IN CORE; 27 SAMPLES

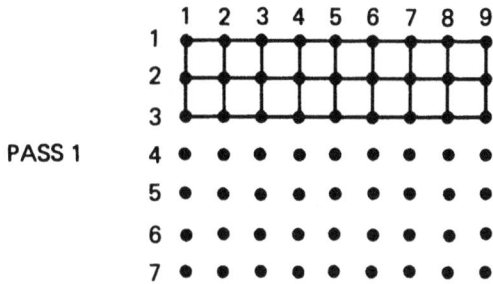

PASS 1 — READ THE FIRST THREE SAMPLES IN EACH RECORD OF FILE 1 INTO THE ROWS OF THE 3 x 9 PARTIAL MATRIX. THEN WRITE THE ROWS ONTO FILE 2.

REWIND FILE 1

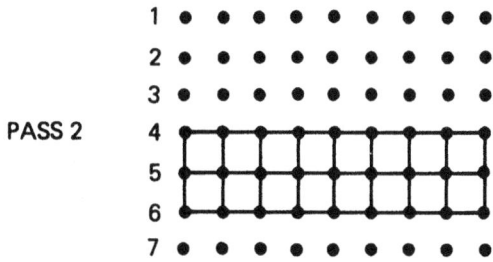

PASS 2 — READ SAMPLES 4, 5, AND 6 IN EACH RECORD OF FILE 1 INTO THE ROWS OF THE 3 x 9 PARTIAL MATRIX. THEN WRITE THE ROWS ONTO FILE 2.

REWIND FILE 1

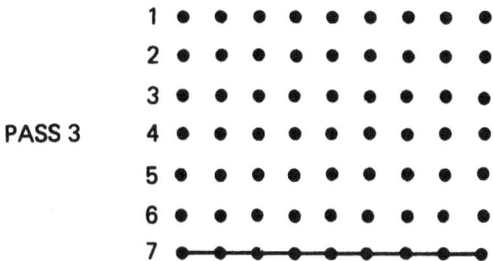

PASS 3 — READ SAMPLE 7 IN EACH RECORD OF FILE 1 INTO A ROW AND WRITE THE ROW ONTO FILE 2.

After we have obtained $Y(\tau_m, \nu_k)$ in range sequence, we can implement the discrete circular convolution

$$Z(\tau_m, \nu_k) = Y(\tau_m, \nu_k) \circledast X_P(\tau_m) \, , \, k = 0, \ldots, N-1 \quad \text{(11.49)}$$

which is also best done indirectly via two FFT's; for our example, one of length $N_\tau = 1000$, the other of length $N_\tau/2 = 500$. (Since we need only every other range sample on output, it is possible to combine samples in pairs after the first FFT, and thus reduce the size of the second FFT by half.) With a CDC6400 computer the computation time for $N_\tau = 1000$ is about $(1000 \times 21 + 500 \times 19) \times 16 \, \mu sec = 490 \, msec*$ for each row corresponding to a Doppler filter sample, and about 65 sec for the N = 128 rows. On output we are interested in the 375 range samples beginning at r_{MIN} and spaced by $\Delta r_{RES} = 2\Delta r$.

If N_τ is not rich in the factors of 2, 3, and 5, the time required to implement Eq. (11.49) could be large. For example, with $N_\tau = 1010$ the computation time is estimated to be almost 10 times as large as that above. To avoid this dilemma we can make the convolution in Eq. (11.49) an ordinary, instead of circular, one. Therefore, we have to repeat the first $N_P = cT_P/2\Delta r = 200$ range samples of $Y(\tau_m, \nu_k)$ at the end of the array as

$$Y(\tau_{m+N_\tau}, \nu_k) = Y(\tau_m, \nu_k) \, , \, m = 0, \ldots, N_p - 1 \quad \text{(11.50)}$$

Then the ordinary discrete convolution is given by

$$Z(\tau_m, \nu_k) = \sum_{i=0}^{N_\tau + N_p - 1} Y(\tau_i, \nu_k) X_P(\tau_m - \tau_i),$$

$$m = N_p, \ldots, N_\tau - 1; \quad k = 0, \ldots, N-1$$

$$\text{(11.51)}$$

---

*The time is based on Nx(Σ of prime factors of N) x 16 μsec on the CDC6400, where N is the length of the FFT. If N is some integral power of 2, then the 16 μsec is reduced to 10 μsec, which is equivalent to (N $\log_2$ N) x 20 μsec. If the above convolution were implemented directly over the 401 samples of the pulse autocorrelation function and the 375 output samples, the computation time would be about 6 sec for each row.

This convolution can also be implemented via 2 FFT's if we add enough zero samples to make the total array length a good number for the FFT. For the example of $N_\tau = 1010$ and $N_P = 200$ we could add 70 more zeros to make the array length 1280; the computation time would be increased by only 30% over the original 1000-point circular convolution.

Thermal Noise

When we simulate the receiver response $Z(\tau, \nu)$ directly on the basis of the linear processing assumption, we must wait until $Z(\tau, \nu)$ is created before we add thermal noise. The reason is that Eq. (11.45) includes the effects of the transmitted signal as well as those of the receiver. Receiver noise is not affected by the transmitted signal.

If we refer to Eq. (5.92), the noise power at the output of the receiver (after pulse compression and Doppler filtering) is given by

$$P_N = N_o E_{PF} \sum_{\ell=0}^{N-1} |\omega_\ell|^2 \quad \text{(11.52)}$$

where N_o is the noise power density, E_{PF} is the energy in the pulse waveform to which the receiver is matched, and $\{\omega_\ell\}$ is the sequence of pulse weights. When we add the thermal noise we must be sure to scale the transmitted pulse waveform $\mu_{PT}(t)$ properly. As we have written Eq. (11.31), the transmitted pulse must be scaled by the transmitted power. If we scale $\mu_{PT}(t)$ to unit peak height, then we can include the peak transmitter power in Eq. (11.31). The scaling of the pulse waveform to which the receiver is matched is arbitrary, as long as the same scaling is used to compute E_{PF} above and in the calculation of $X_P(\tau)$.

The thermal noise is added in the form of random phasors to the samples $Z(\tau_m, \nu_k)$. For our example we have to generate 375 x 128 = 48,000 random phasors, or 96,000 Gaussian random phasor components. Using the fast methods in Section 9.2 we can generate a Gaussian random

Table 11.2/Summary of Steps in Generating and Processing the Video Signal for One Transmission

Step	Matrix Size		Computation Time (sec)
	Range	Doppler	
1. Generate and Transform Ground Clutter	1000	256	500
2. Create Complex Video Signal	1000	256	15
3. Doppler Convolution	1000	128	100
4. Transpose Matrix on Auxiliary File	1000	128	80
5. Range Convolution	375	128	65
6. Add Thermal Noise	375	128	2
		TOTAL	~750 sec

number in about 14 μsec on the CDC6400 computer. Thus, for the 96,000 random numbers, we need about 2 sec, including the time for scaling.

In Table 11.2 we have summarized the steps required to generate and process the ground clutter video signal for an airborne search radar. We see that the computation time is dominated by Step 1, the generation of the ground clutter RCS and the transformation into radar space. If fact, as we pointed out, this part of the computation is dominated primarily by the antenna pattern calculations. But once we have created the RCS or power matrix in radar coordinates, it is possible to replicate the statistics in Step 2, the generation of the complex video signal, to obtain more output samples without reimplementing Step 1. If we performed this process 10 times whenever we execute Step 1, the average computation time for each transmission would go down from about 750 sec to 300 sec.

For each transmission we have simulated 128 Doppler filter outputs at each of 375 range samples. The total number of samples is 375 x 128 = 48,000. However, some of these cells must be ignored (blanked) because they contain mainbeam clutter and the intense altitude return.

For our example we can estimate the Doppler width of the mainbeam clutter as

$$\Delta \nu_{MB} = \frac{2V}{\lambda} \sin\theta \; \Delta\theta_{MB} \tag{11.53}$$

At broadside ($\theta = 90°$), where the Doppler width is largest, the ratio of Eq. (11.53) to the Doppler resolution is

$$N_{BLANK} = \frac{\Delta\nu_{MB}}{f_r/N} = N \frac{2V}{\lambda f_r} \Delta\theta_{MB} \tag{11.54}$$

For a one-way half-power width of the mainbeam of 2°, the two-way 40-dB width is about 2.5 times as large, or 5° (.0873 rad).

In our example, N = 128 and 2V/λf_r = 1.6. Therefore the number of filters that have to be blanked is 128 x 1.6 x .0873 = 17.9 or 18 filters. As we scan off-broadside, the number of filters containing mainbeam clutter is reduced by $\sin\theta$. Thus, at $\theta = 45°$, we have to blank only 13 filters. Not counting the altitude return, the number of false alarm opportunities is thus 375 x 110 = 41,250 at broadside. This figure is reduced by about 100 samples when the altitude return is blanked.

Chapter 12

Real-Time Radar Signal Simulation

by G.E. Pollon and J.F. Walker

Malibu Research Associates
Technology Service Corporation
Santa Monica, California

The real-time generation of radar signals for use in actual radar processing represents the final stage in the application of simulation technology to its ultimate objective — the testing of radar hardware and its associated system. In this chapter we introduce the principles of real-time radar signal simulation and provide a discussion of the major design and implementation problems. The pacing considerations in this type of simulation, compared to one which is fully computer implemented, are the speed requirements of the on-line signal generation and the interactive nature of the radar with the state of the environment. The relationships between the critical off-line and on-line components of the simulation essentially are dictated by such considerations as they apply to the object radar and its operating environment. We will provide examples of specific approaches to implementing the basic simulation functions, discussed in the context of a pulsed ground based radar as the object system. However, except where otherwise noted, the approaches given are far more universal.

It is important to note that the feasibility of a real-time radar signal simulation is based on the nature of the received radar signal which, for all practical purposes, consists of delayed and Doppler shifted replicas of the transmitted waveform. The entire information content of the environment exists as modulations of the transmitted waveform.

The applications of a well conceived real-time radar signal simulation extend through all phases of the development, test, and operational use of the object radar. Such a simulation can and has been used successfully as a test bed for radar data processing development and debugging; as a tool for testing and evaluating a completed prototype system; for operator training; and for exercising the radar as an element in a total system — all being accomplished without the benefit of live measured data on targets and clutter. The superposition of terrain and weather clutter on true target signals and the modeling of targets under actual clutter conditions are additional simulation uses. Further applications include system maintenance and production unit testing under operating conditions that are more complex than those achievable with simple test signals.

A real-time environment simulation is able to produce test conditions that include the entire target and clutter environment behavior; thus a single simulated test signal injected into the front of the radar will exercise properly all radar signal processing functions — including MTI, CFAR, and detection — and all data processing functions — including discrimination, tracking, data handling, and display. Operator functions can be exercised as well. Moreover, since the radar output can act as an input to subsequent system operations, the modeled signal can be a stimulant or exercisor for an entire system. We show how the simulation interacts with the various radar processing functions in Fig. 12.1.

12.1/Components of the Real-Time Signal Simulation

We define a real-time simulation as a system whose output is a signal stream suitable for injecting into the front end of an actual radar receiver at RF, IF, or video. The nature and data rate of the output implies that some special purpose digital or analog hardware is used to accom-

Figure 12.1/Real Time Environment Signal Simulation
for System Test

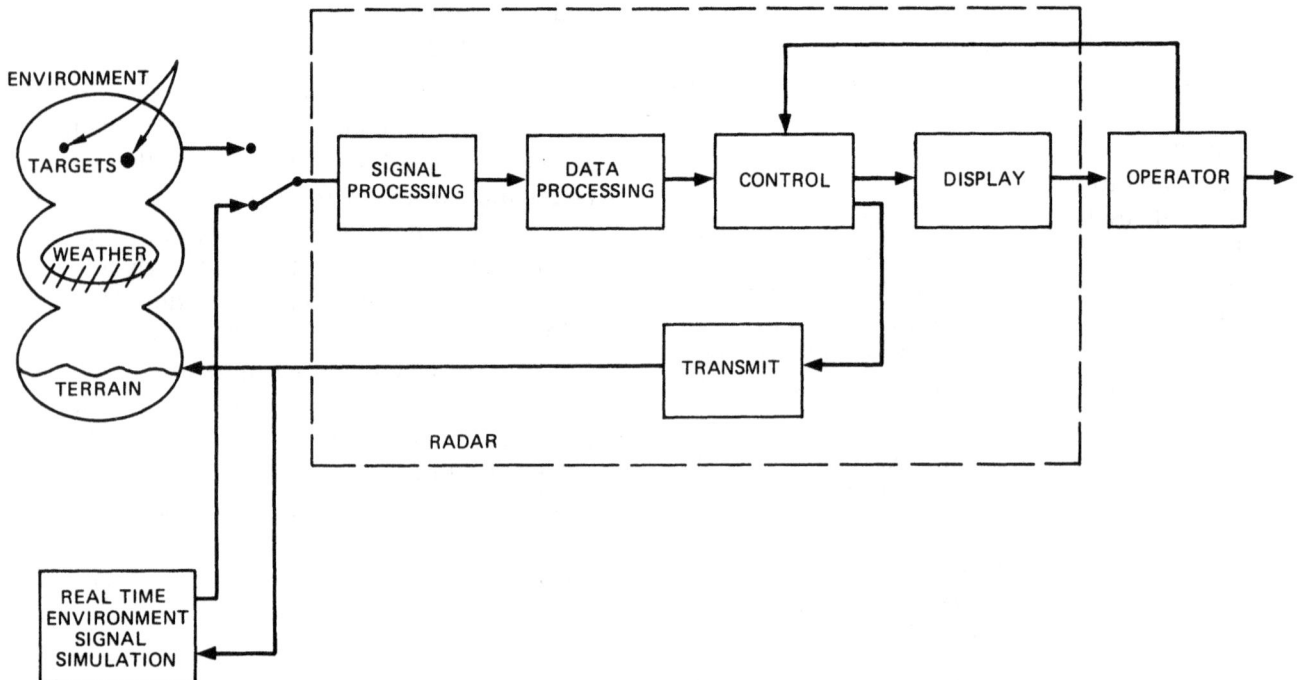

plish the output interface with the radar. The simulation aspect of the process indicates the use of a computer for describing the radar environment — the location of targets, their trajectories, the location of clutter, etc. — though the use of actual measured data is not precluded. The construction of a data base to describe the radar environment usually takes place in a large-scale general purpose computer that is off-line. A real-time radar signal simulation may be defined broadly as a system or process that starts with a general purpose computer implementation of the environment and ends with the production of real-time signals injected into the front end of a radar receiver or processor, including at least some special purpose interface hardware.

For a simple radar, these components may be the only ones in the simulation. For a radar

that does not interact either with an operator or with the state of the received signal, it may be possible to precompute and output the simulated signals directly to a simple storage and timing interface unit. But in the more interesting and general case of an interactive radar system, especially one with a high signal bandwidth (> 100 kHz), the general purpose computer cannot do all of the signal computation — intermediate processors are necessary.

In Fig. 12.2 we show the more general form of a real-time signal simulation, consisting of five distinct processing operations, each based on a different processor or device:

1) Data Base Generation — to be done off-line
with general purpose
computer

*2) Bulk Data Storage
and Transfer* — via magnetic tape

3) On-Line Preprocessing — with a minicomputer

4) Signal Generation — with special-purpose digital hardware

5) Real-Time Output — D/A conversion and and analog processing

The size and importance of any one of these operations varies considerably, depending upon the application. One or more may be combined in some cases, but the functions provided by each usually must be present in one form or another. We now discuss these operations in detail.

Data Base Generation

The data base generation is the basic simulation process by which descriptions of the target and clutter environment are provided to the on-line processing; it is based on target and clutter environment models and the known characteristics of the radar. To ease the burden of real-time computation it is desirable to focus time and effort on the preparation of the data base. However, this preparation is limited by a fundamental lack of *a priori* knowledge of the sequence of events in an interactive radar. In addition, the amount of off-line data preparation is limited practically by the volume of data and the speed transfer problem between off- and on-line portions of the simulation; the latter limitation can be particularly severe in the generation of clutter signals.

The central and most appealing feature of using a GP computer to construct the data base is the potential to use realistic models and actual data for target and clutter representation, and the ability to perform complex manipulations of them. In the end, those operations establish the ultimate credibility of the simulation process.

Bulk Data Storage and Transfer

It is most economical to generate as much detailed data off-line as possible, but this action is limited by the storage and transfer of data. There are several devices available for those purposes, though, including magnetic tape, disc, and drum. The latter two have the advantage of much higher speed than the magnetic tape unit; but beyond a certain data transfer rate, this advantage ceases as data storage volume becomes the limiting factor.

Figure 12.2/Elements of a Real-Time Radar Signal Simulation

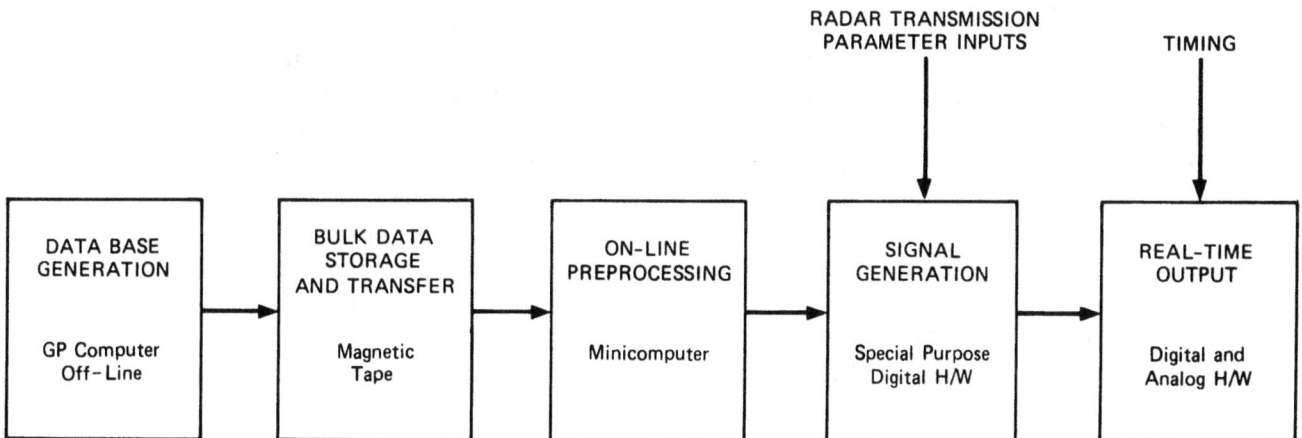

An example illustrates the nature of this problem. A 2400-ft magnetic tape contains approximately 135 million bits of data. At 75 in/sec, the data transfer rate is 3.6×10^5 bits/sec and the tape will run for approximately six minutes (about the minimum that is of interest for most interactive simulation scenarios). The higher data rates achievable with other devices lead to shorter run times. The data capacity and cost of magnetic tape systems, in comparison to other devices, make them the natural choice for a real-time radar signal simulation. However, the resulting constraint put on the transfer rate between the off-line and real-time operating portions of the simulation becomes a dominant factor in the design of the system.

An alternate approach to an intermediate bulk storage device is the use of an on-line GP computer to perform the data base generation in near real-time. However, even when it is possible to dedicate a large-scale computer to one application, it is not possible to perform the required computations for reasonably sophisticated target and clutter models at a rate equivalent to the tape data transfer rate. For example, the simulation in Chapter 10 would be about 3 orders of magnitude slower than real time on a CDC6400 computer.

On-Line Preprocessing

On-line preprocessing includes those functions that interface with the data base and the real-time radar signal generation. Included are data read and format operations, pre-run data selection, calculation, initialization, and operator control interface; some (near) real-time data manipulation may or may not be included in addition to data transfer. The basic purpose of preprocessing is to put the data into a form as convenient for the signal generator as possible. The distinction between the preprocessor and signal generator lies in the fact that the latter cannot perform its functions until the radar beam

steering and waveform commands are available, whereas the former is free to precompute its operations in non-real time. Since any precomputed functions can be performed in the off-line GP computer, the preprocessor is redundant for all but data transfer operations. However, it is useful for performing some functions in order to reduce the transfer load; also, since the preprocessor is an on-line component, it can perform data manipulation or selection in response to operator control inputs. But flexibility via on-line operator control, either just prior to or during a simulation run, is usually a prime objective; thus, the preprocessor is easily justified as a separate simulation unit.

A minicomputer is the natural means of accomplishing the preprocessor operations. It can readily perform interface functions among a magnetic tape unit, the signal generator, and operator commands; it is well suited for pre-run data manipulation and on-line calculation. The choice of a minicomputer, rather than simpler special purpose hardware, depends primarily on the degree of desired operator control and the extent to which such controls impact on the data calculation.

Signal Generation

The preprocessor provides inputs to the signal generator, the heart of the real-time aspect of the simulation. The function of the signal generator is the (near) real-time generation of target and clutter signals based on the instantaneous radar transmission parameters (beam and waveform) and the current target and clutter environment state (as provided by the data base description). The interactivity between the real-time simulation and the radar takes place in the signal generator; it is the prime component that is responsive to the instantaneous radar transmission. The structure and speed of the entire simulation is based largely on the parameters of the interface between the simulation and the radar, since they determine both the amount of off-line computation that can be performed and the complexity and speed of the on-line signal generator operations.

The operations involved are the computation of the antenna beam pattern, or the spatial filtering of the environment, and the application of the transmitted waveform to the environment. The former usually requires only a knowledge of the instantaneous beam pointing angle for a fixed beamshape antenna, but it may also include additional spatial filtering parameters as, for example, with an adaptive array antenna. A case of particular interest is the formation of antenna monopulse patterns for tracking radars.

With a pulsed radar, in all cases of practical interest, the signal generator response to the transmitted waveform may be considered as a delayed and Doppler shifted replica of the transmitted waveform applied to each point in the target and clutter environment. It is assumed that the instantaneous waveform bandwidth is not sufficiently wide to result in either Doppler or spatial beamshape smearing. Either case could be accommodated, but the simulation implementation would be greatly complicated.

It is convenient to separate the signal generation into the computation of both the initial amplitude and phase of a pulse and the production of the pulse waveform. The former operation represents the environmental response to an ideal or infinitesimally narrow pulse transmission; the latter utilizes the finite resolution and coding properties of the transmitted pulse. The interactive signal generator operations take place on a time scale corresponding to the radar interpulse period and beam dwell time. On the other hand, the pulse formation that occurs in the real-time output portion of the simulation in Fig. 12.2 takes place on a time scale corresponding to the signal bandwidth.

Real Time Output

Generation of the output signal is the final step. It must be done in real time in synchronization with the real-time clock. Each target is placed in its proper time slot after the transmission trigger, the pulse coding and waveform shape, based on an ideal pulse, is applied to each signal generated.

The simulation output element also accomplishes the interface with the radar input requirements, which may involve changeover to the carrier or intermediate radar frequency through appropriate digital to analog conversion and modulation circuitry. If the radar input is to take place at video, then the output must reflect any signal processing or filtering that takes place prior to the point where the signal is injected.

12.2/Relationship of Radar to Its Environment

The radar and its environment are dynamic — the former by means of the changing position of its antenna beam and the timing of its waveform; the latter through the motion of scatterers and their changing scattering properties. The radar, with its antenna beam and waveform, is able to sample the state of the environment; moreover, it can react to that state by changing certain parameters or its mode of operation. This property is fundamental in a real-time simulation.

Time Scales of Events

We can define four time scales that describe the state of the environment, the state of the radar, and the interaction of each, as shown in Table 12.1. Depending upon the radar parameters, the time scales of particular events may overlap; for a considerable range of parameters, though, the time scales are usually distinct.

Table 12.1/Time Scales of Radar-Environment Events

Time Scale	Events
Minutes or hours	Definition of Environment
	Radar location
	Ground clutter
	Rainfall rate
	Wind speed and direction
	Target properties
Tenths of seconds	Physical Motions
	Radar platform motion
	Target motion
Milliseconds	Time on Target
	Beam dwell time
	Pulse repetition period
	Clutter fluctuations
Microseconds	Video Bandwidth

On the slowest time scale we describe what the underlying environment is — the regions of clutter, the rainfall rate, the wind properties, and the target conditions and prescribed trajectories. These conditions change only very slowly and can usually be treated as being constant and defining a single scenario. The physical motion of the environment (principally targets and the radar platform) takes place on the next slowest time scale, which is of the order of tenths of seconds, ultimately depending on radar measurement accuracies and target velocities. From the viewpoint of signal simulation, the most basic time scale of the radar-environment interaction is that of the interpulse period and the antenna

beam dwell (time on target). The latter may last several interpulse periods, but is still on the same time scale, typically milliseconds. The time scale of the PRF and beam dwell establishes the timing and computation rate to which the real-time signal simulation must conform. The final and shortest time scale is that established by the radar signal bandwidth, typically of the order of a microsecond or less. The most difficult simulation computation load can be imposed by this time scale; fortunately, the problem doesn't arise under reasonably low target density conditions, i.e., when only a relatively small number of radar range resolution cells are occupied by targets. The significance of the different time scales is most evident when selecting the approaches to be used in the implementation of the real-time signal simulation.

Radar-Simulation Interactivity

The ability of the radar to react to the state of the environment is the second major consideration in establishing the structure of the real-time signal simulation process; in Fig. 12.3, we illustrate the form of that interaction. The radar acts on the target and clutter through spatial and time stimulation, i.e., through the antenna pattern and the transmitted waveform. The returns from the target and clutter to the radar processing are functions of the scanning beam pattern and the transmitted waveform. The simulation is defined as *reactive* if the radar beam positioning and/or waveform selection are determined by the nature the signal returns; it is *nonreactive* if the beam positioning and waveform proceed in a preestablished pattern, independent of both the signals received and the results of subsequent processing.

A common example of a nonreactive case versus a reactive one is that of mechanically scanned systems and phased arrays. In the former the scan of the beam is predetermined by the physical rotation of the antenna, whereas in the latter the positioning of the beam may change, depending upon the result of data processing on the received signal. For example, if a target detection occurs in a phased array radar, the search may be interrupted to obtain track data; interactivity with the transmitted waveform may also occur by changing the PRF or signal bandwidth.

The predominant influence of interactivity on the structure of the real-time signal simulation is focused in both the data flow and timing aspects of the simulation. The lack of prior knowledge as to the environmental stimulus (beam and waveform) means that the data base of the simulation, describing both targets and clutter, must be broad enough to encompass the range of likely stimuli. Similarly, the actual generation of the environment response signal can only be initiated after the beam position and waveform have been designated by the radar.

The presence of interactivity, in combination with the time scales on which the major functions must be accomplished, effectively defines

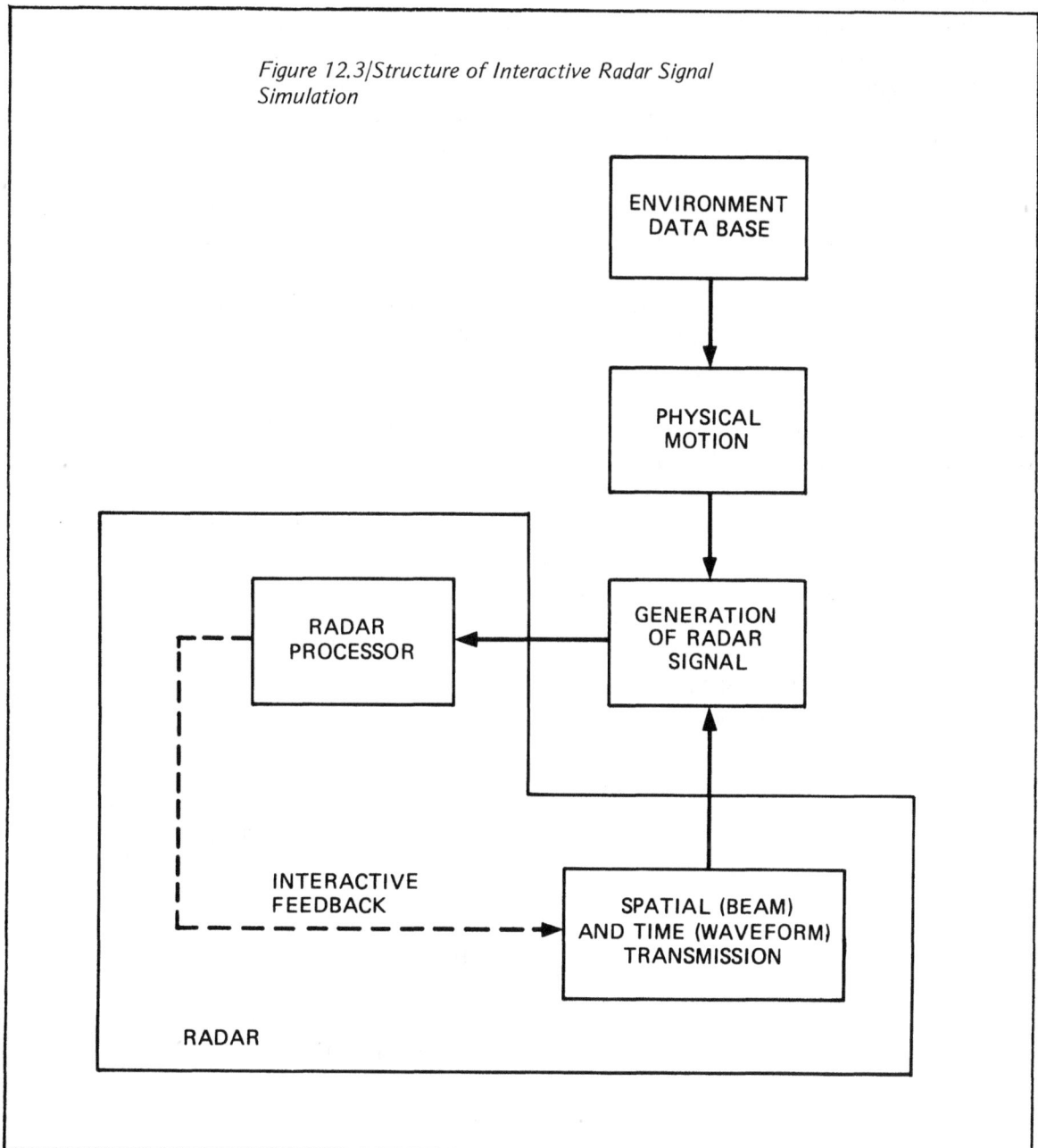

Figure 12.3/Structure of Interactive Radar Signal Simulation

which parts of the simulation must be carried out in real time; which must be accomplished in near-real time; and those that may be pregenerated off-line.

Radar Interface

The radar-to-environment interface is specified by the antenna beam and transmission waveform parameters. This information must be intercepted by the real-time simulation and used at the moment of transmission to appropriately generate the simulated signal returns. The state of the antenna beam is usually indicated by an instantaneous azimuth and elevation pointing angle. The beamshape is assumed to be known *a priori*, although it may change as a deterministic function of the scan angle. In the case of a monopulse antenna, three antenna patterns and three corresponding received signals may be involved.

The transmitted waveform is usually specified by a pulse transmission time (trigger) and a transmitted frequency. If the radar has multiple transmitted pulse waveforms, then an indicator of which waveform is in use must also be provided to the real-time simulation with each transmission.

It is usually desirable for the radar clock to be used by the simulation in order to achieve stable pulse signal synchronization. Similarly, if the object radar receiver performs sequenced signal processing operations, such as time multiplexing of three antenna monopulse signals, then appropriate timing gates must be made available to the simulation.

12.3/Data Handling Versus Computation

A reoccurring problem in the design of a real-time signal simulation is the tradeoff between the real-time transfer of data computed off-line and the computation load to be performed by the real-time components. The latter represents the greatest cost element of the simulation; consequently, it is generally desirable to do all computation off-line up to the point where the radar interacts with the environment. Depending upon the complexity of the simulated environment and the number of radar parameters, it is usually not possible to transfer a complete data base to the real-time portion of the simulation at the required rate. In such cases, a more limited form of the data can be computed off-line, but even more processing by the real-time system is then required. The tradeoff that one encounters is generally a question of updating discrete events, such as the coordinates defining a target trajectory, at a high rate for direct use on-line versus a lower rate of updating coupled with on-line interpolation or filtering of the data. The accuracy requirements are basic to this tradeoff, as is the dynamic nature of the environment.

A tradeoff between data handling and computation load also occurs within the on-line processing system, or alternately within the off-line generation system. This tradeoff is that between bulk data storage, in a finely incremented table lookup form, versus the computation of functions. It can have significant cost impact within the real-time system, and depends once again on the overall simulation accuracy required for the radar application.

Figure 12.4/Various Methods of Sampling and Processing Continuous Data

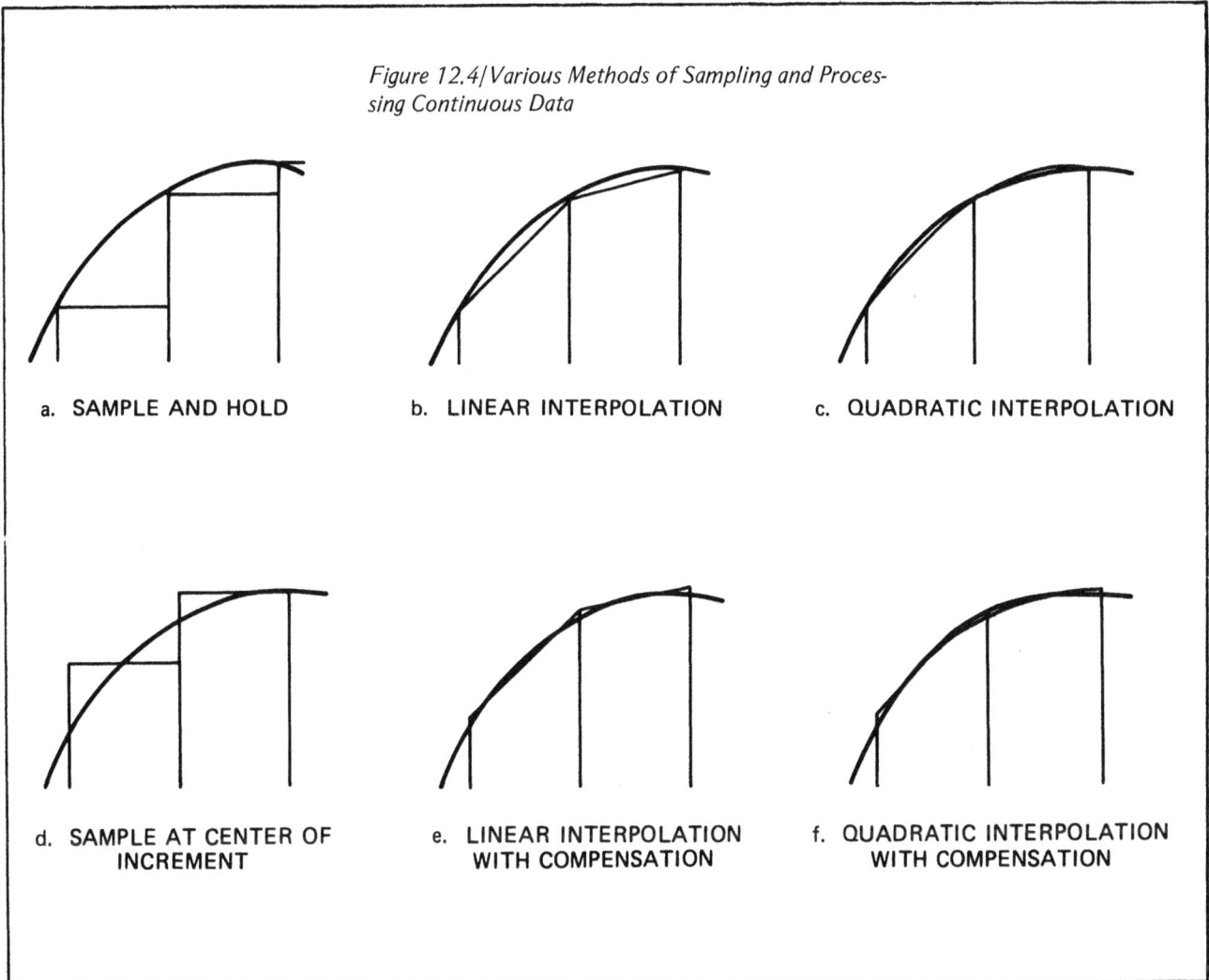

a. SAMPLE AND HOLD   b. LINEAR INTERPOLATION   c. QUADRATIC INTERPOLATION

d. SAMPLE AT CENTER OF   e. LINEAR INTERPOLATION   f. QUADRATIC INTERPOLATION
   INCREMENT   WITH COMPENSATION   WITH COMPENSATION

Deterministic Variables

Deterministic data associated with target trajectories, antenna patterns, and library functions, such as sine and cosine, must be sampled. There are various ways to sample and process such data, as is illustrated in Fig. 12.4. Sample and hold, linear interpolation, and quadratic interpolation are straightforward. To reduce the errors associated with sample and hold, we should sample at the center of each increment, as shown in (d). As we discussed in Chapter 9, with linear interpolation we can adjust or compensate each

data sample to remove the mean error within each increment* as shown in Fig. 12.4e. Compensating for quadratic interpolation, as in (f), can be accomplished in a similar manner. In Table 12.2 we summarize the peak and rms errors in terms of the sampling increment Δt and

*To minimize the *peak* error after compensation, the correction should be 1/2 of the error prior to compensation. The resulting rms error will be reduced to 47% of its previous value. On the other hand, to minimize the *rms* error the correction should be 2/3 of the peak error prior to compensation, as we discussed in Chapter 9. The resulting peak error will be 33% larger, but the rms error will be 41% of its prior value.

the data derivatives. The requirements on sampling errors associated with deterministic data are constrained by the ultimate measurement precision of the radar. (An exception we will discuss is with the computation of sine and cosine from phase.)

Table 12.2/Interpolation Errors for Deterministic Variables

	Peak	RMS
Sample and Hold	$1.0 \, \dot{x} \, \Delta t$	$0.58 \, \dot{x} \, \Delta t$
Sample at Center of Increment	$0.5 \, \dot{x} \, \Delta t$	$0.29 \, \dot{x} \, \Delta t$
Linear Interpolation	$0.125 \, \ddot{x} \, (\Delta t)^2$	$0.091 \, \ddot{x} \, (\Delta t)^2$
Linear Interpolation (Compensated)	$0.063 \, \ddot{x} \, (\Delta t)^2$	$0.037 \, \ddot{x} \, (\Delta t)^2$
Quadratic Interpolation	$0.064 \, \dddot{x} \, (\Delta t)^3$	$0.046 \, \dddot{x} \, (\Delta t)^3$
Quadratic Interpolation (Compensated)	$0.042 \, \dddot{x} \, (\Delta t)^3$	$0.025 \, \dddot{x} \, (\Delta t)^3$

As we can see from Fig. 12.4, errors can be greatly reduced or the sampling increment requirement greatly relaxed, by the use of linear or quadratic interpolation. The cost is an increased computation load during the on-line portion of the simulation. Various compromises are possible. For example, we can provide the on-line processor with data samples and the slope values to be used in interpolation, thereby reducing the demand somewhat. More importantly, though, both range and range rate data could be provided to the on-line processor in order to avoid the higher sampling rate required to differentiate range data samples on-line.

Stochastic Variables

In the case of simulating stochastic variables, there arises a similar problem between data handling and the on-line computation load; however, the criteria for accuracy are different than those for deterministic variables. For random

processes we are concerned primarily with spectral properties. Any errors introduced by sampling and subsequent processing must be evaluated in terms of their effect on the spectrum.

In Chapter 8 an analysis is given of the errors associated with interpolating random processes; the results are summarized in Fig. 8.10. In real-time an additional complication arises when random processes are frequency-shifted from dc, such as wind-driven rain clutter as seen by a ground based radar or ground clutter as seen by an airborne radar. In a non-real time simulation the frequency offset can be applied after interpolation by a simple phasor multiplication; in real time, though, this action may not be possible because of constraints on the computation load. In such cases frequency shifting must be applied to the data base *before* interpolation. The residual interpolation error, is affected by such an application as we see from Eq. (8.57), if we offset one of the terms within the integrand. If both the width of the spectrum and the spectral offset are small compared to f'_r, the sampling frequency when interpolation is implemented, we can rewrite the spectral residue in Eq. (8.59) with an offset as

$$R_n(f_o) = (\Delta t')^{2n} \int |X'(f)|^2 (f - f_o)^{2n} df \qquad (12.1)$$

where f_o is the frequency shift from dc and $\Delta t' = 1/f'_r$. Thus for any symmetrical spectrum we can write

Sample & Hold (n=1)

$$R_1(f_o) = R_1 + (f_o \Delta t')^2$$

Linear (n=2)

$$R_2(f_o) = R_2 + 6(f_o \Delta t')^2 R_1 + (f_o \Delta t')^4 \qquad (12.2)$$

Quadratic (n=3)

$$R_3(f_o) = R_3 + 15(f_o \Delta t')^2 R_2 + 15(f_o \Delta t')^4 R_1 + (f_o \Delta t')^6$$

where R_n are the spectral residues given in Section 8.5 with no spectral offset (we assume that the spectral power is normalized so that $R_o = 1$). For the Gaussian shaped spectrum we can use Eq. (8.61) to write

Sample & Hold

$$R_1(f_o) = R_1 [1+5.55 (f_o/f_{3dB})^2] \qquad (12.3)$$

Linear

$$R_2(f_o) = R_2 [1+11.09(f_o/f_{3dB})^2 + 10.25(f_o/f_{3dB})^4]$$

Quadratic

$$R_3(f_o) = R_3 [1+16.63(f_o/f_{3dB})^2 + 30.73(f_o/f_{3dB})^4 + 11.36(f_o/f_{3dB})^6]$$

In Fig. 12.5 we show the quantity within the brackets for the three cases. If we add the specified correction to the curves in Fig. 8.10 we obtain the residual power for the offset spectrum. For the case where $f_o \gtrsim 1.5 f_{3dB}$, the last term in each expression in Eq. (12.2) dominates and we can write

$$R_n(f_o) = (f_o/f'_r)^{2n} \qquad (12.4)$$

where f'_r is the sampling rate prior to interpolation.

Usually ground clutter is much more intense than rain clutter. For a ground based radar the former is centered at dc; we can use Fig. 8.10 to determine the sampling rate necessary to achieve a low given value of residual spectral power. For example, the requirement might be

-60 dB for ground clutter, in which case the sampling rate for linear interpolation would have to be about $f'_r = 18 f_{3dB}$. Rain clutter is usually offset from dc, but since it is not as intense as ground clutter we could probably tolerate -40 dB of residual spectral power. If we use Eq. (12.4) for linear interpolation (n = 2), we obtain $f'_r = 10 f_o$. Thus a sampling rate of $f'_r = 1000$ Hz (this value will be assumed in the example of Section 12.5) accommodates maximum values of $f_{3dB} = 55$ Hz for ground clutter and $f_o = 100$ Hz for rain clutter.

12.4/Simulation Models

The validity of the signal simulation process is ultimately dependent on the realism and credibility of the underlying target and clutter models. Since they are the backbone of the real-time simulation, an implementation which sacrifices fundamental target and clutter characteristics for the sake of expediency is of questionable value. Since target and clutter modeling has already been discussed in Chapter 3 and since numerous references are available on the subject, we shall examine those characteristics of targets and clutter which impact on the structure of the real-time simulation.

Targets

A point target is defined by its RCS variation as a function of time and by its trajectory, which for a ground based radar can be in the radar coordinates of range, azimuth, elevation, and range rate. For our purposes the trajectories are considered to be deterministic and, as such, are not subject to modeling; they are configured according to the test scenario. RCS fluctuations, however, are subject to modeling, since this characteristic is usually incidental to the test scenario.

Figure 12.5/Correction for Curves in Figure 8.10 for Offset Spectrum.

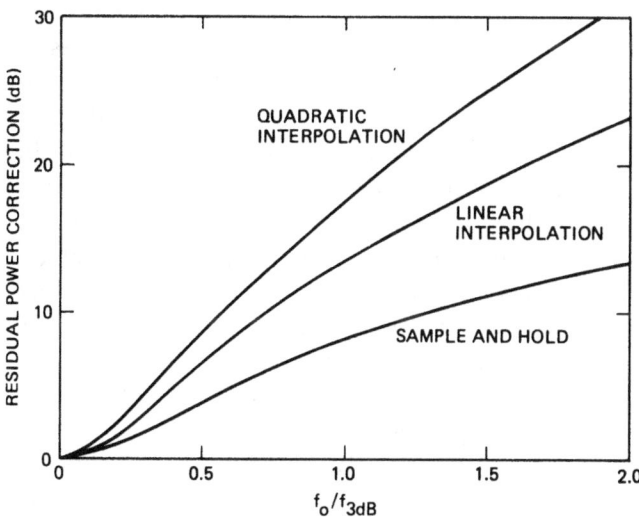

As discussed in Chapter 3 the RCS fluctuation may be obtained for the simulation by means of an appropriate statistical model such as the Rayleigh, chi-square, or log-normal models, or via direct RCS data obtained by measurement or computation. The choice depends primarily upon the time history of the target aspect angle, the target size, and the availability of a comprehensive set of data. In cases where the history is well behaved and can be deterministically established, the most accurate approach is to use available measured or computed RCS data; in general we require dual polarization and phase data. When a degree of aspect angle uncertainty is present that is comparable to the target lobe width of about $\lambda/2L$, then a statistical model must be relied upon. A hybrid model, in which the statistical fluctuation parameters are taken as slowly varying functions of aspect angle, is frequently the most feasible approach.

How the fluctuations are modeled primarily affects the off-line data base generation. Sample target RCS values, whether generated by a statistical or deterministic approach, constitute a varying target parameter which is a continuously updated component of the data base. The structure of the simulation is affected only by the requirement imposed by the RCS update rate.

RCS time fluctuations represent a first order radar-independent property. Radars that employ frequency diversity pose a more complex problem in modeling target fluctuations because of an interdependence between fluctuation and frequency. If the radar frequency steps are sufficiently large, independent fluctuations may be assumed and the modeling problem is simplified. For smaller step sizes, however, a correlated statistical fluctuation model must be employed. In the case of slender axisymmetric bodies the relationship described by Eq. (3.7) may be used to obtain RCS data for small changes in frequency. If it is not known *a priori* what frequency will be used by the radar on a given transmission,

a set of RCS data or suitable frequency sensitive coefficients must be provided to the on-line simulation as a part of the data base.

Even though such objects are not actually targets, the point clutter associated with occasional large man-made objects can be modeled as a collection of targets for simulation purposes, provided the number of objects is few. This approach applies to false targets as well.

Distributed Clutter

Models for distributed ground and rain clutter have a different character than do those encountered for target signals. Targets essentially are defined by their discrete low density nature and a trajectory in real space. Clutter, on the other hand, is characterized by its continuous high density nature, i.e., a high percentage of radar range-azimuth resolution cells contain clutter signals. The primary properties of distributed clutter that the simulation must encompass are fluctuation (mainly through its Doppler spectrum) and its spatial distribution.

The nonhomogeneous distribution of the ground clutter backscatter coefficient, σ_o, is a result of the random nature of terrain. Over a widespread area, variations occur in local grazing angle, vegetation type and density, ground moisture content, and other factors that affect the local backscatter coefficient. Large magnitude changes can result from unique localized terrain conditions and from line-of-sight shadowing of portions of the surface. This irregular nature of the backscatter coefficient may be modeled statistically, as was done in Chapter 10, with a log-normal distribution commonly used. In Table 12.3 we give some representative values for the median and standard deviation of the statistical backscatter coefficient.

In Chapter 10 we described an example of generating a ground clutter map in radar coordinates; it is practical to perform this operation off-line as a part of the data base. Rain clutter may also be described in terms of a radar map, incorporating nonhomogeneities in the rainfall rate, weather fronts, etc.

Table 12.3/Log-Normal Parameters for Typical Terrains (Referenced to S-Band and 5° Angle of Incidence; after [1])

	Median Value of σ_0	Log-Normal Standard Deviation
Desert	-30 dB	10 dB
Woodlands	-26 dB	8 dB
Farmland or Clear Flatland	-24 dB	8 dB
Heavy Vegetation	-22 dB	8 dB
City	-15 dB	10 dB

It is common to assume a Gaussian spectral shape for ground clutter. Other data, however, indicates a spectrum of the form [Ref. 2].

$$H(f) = \frac{1}{1 + |f/f_o|^3} \qquad (12.5)$$

This latter model predicts high frequency tails of a much greater magnitude than those that would result from a Gaussian spectral shape. The appropriate model may well depend upon the radar application. As seen from a moving platform, clutter may assume a variety of spectral shapes depending upon the antenna beam pattern, scan angle, and geometry. The implication of these variations in the ground clutter spectrum is that the simulation structure should be capable of generating clutter with various spectral shapes.

The spectrum of rain clutter is also modeled generally as a Gaussian shape [Ref. 3]. For most radar applications that model is adequate; however, in some cases, depending upon the shape of the weather patterns or the geometry, non-Gaussian and skewed rain clutter spectra may

result. Again, the simulation should be capable of accommodating multiple clutter spectral shapes.

Sky Clutter

A unique form of clutter frequently encountered by high sensitivity radars used in surveillance of the lower atmosphere, and one that impacts significantly on the real-time simulation structure, is sky clutter. Commonly known as a radar angel phenomenon, it usually is attributed to migratory bird activity. Such clutter appears as discrete detections to the radar and can be a major cause of false alarms and processor overload. The bearing on the simulation lies in the fact that this type of clutter must be treated as collections of discrete objects, each possessing velocity and fluctuation characteristics. In Table 12.4 we give examples of some of the more important model characteristics of sky clutter (see Refs. 4-8). In a real-time simulation a realistic approach is to generate a population of radar angels, according to the distributions in Table 12.4, and to provide discrete target parameter updates to the on-line simulation as a part of the data base. Because of their low speed, these sky clutter targets can be updated at a relatively low rate.

Table 12.4/Typical Sky Clutter Model Characteristics

Average RCS Distribution	Log-Normal Median RCS = -27 dBsm Log-Deviation = 6 dB (at S-Band [4])
Density	0.3/km², 50th percentile [4]
Velocity Distribution (m/s)	$e^{-(V-10)/5}$, $V \geq 10$ m/s
Altitude Distribution (meters)	$e^{-(h/750)}$ [5]
Fluctuation	Log-Normal [6]

12.5/Data Base Generation

The data base that is accessed during the real-time portion of the simulation consists of a time sequence of *snapshots* of the radar environment. Each snapshot is essentially a list of parameters that describe each target and the state of the clutter, and a set of random samples to be used in the generation of the clutter signals. These snapshots are scaled and converted to make their formats compatible with the simulation on-line hardware. They are stored in time sequence on tape or some other storage medium.

The methods used for the generation of a simulation data base by a GP computer are highly dependent upon the specific nature of the environment to be generated and, to a lesser extent, by the architecture of the simulation hardware. However, irrespective of the particular objectives or constraints of a given simulation system, we can list the general requirements for generating a data base. First of all, we need a description of the targets and their trajectories, which can be in the form of a set of equations or tables; we also need a set of parameters or tables that describe clutter. Secondly, we need a management program or a scheduler that keeps track of the time sequence of events. Thirdly, we need programs that generate the actual snapshots. In these programs we can implement the equations of motion at an instant of time or we can interpolate within tables. Finally, we need to convert or reformat the data to make them compatible with the on-line simulation hardware.

Other functions can be performed as part of the process of generating a data base. For example, it is very desirable to have some means of validating data. Two of the most useful methods are to plot or otherwise display the data, and to implement various statistical tests. With either, it is possible to spot many discrepancies that might otherwise go undetected.

Data Generation

In an actual design we probably need several different types of data generators, each contributing to a given snapshot of simulation data. For example, there will be different types of targets, some of more interest than others. In some cases we want to use measured or computed data to describe certain target properties; in others, statistical data suffices. And distributed clutter is distinctly different than discrete clutter.

The generation of target data can be costly, as illustrated by the following example. Suppose we wish to generate snapshots of the target range, range rate, azimuth, elevation, and RCS every 0.05 sec. We assume that the mission will last 6 min, which means that there will be 7200 data points or snapshots to be generated. If 1 msec of computer time were required for each data point, it would take 7.2 sec to generate the complete mission for one target. On a CDC6400 computer, this procedure is likely to cost $1.00, which is not much for one target. But it would be significant if we had to repeat the computation for 100 targets. Clearly, with so many targets, it is not desirable to exceed the 1 msec per data point, and it would be highly desirable to reduce that time somewhat. It turns out that a straightforward implementation of the equations of motion in FORTRAN on a CDC6400 requires about 1 msec per data point. Thus there is some motivation to implement a faster approach. One way would be to use assembly language instead of FORTRAN, but that approach is not recommended except in very limited instances. Another method would be to precompute the trajectories at coarse increments in time and store them in tables; interpolation would then be used to generate the data at finer increments. Here we are trading more core storage for a faster computation. A simple approach would be to implement the fast algorithms in Chapter 9; the computation time would be reduced to about 1/3 of its previous value.

An Example

Suppose we wish to generate a tape that contains the scenario described in Table 12.5. During real time this tape must transfer data at the rate of 10,000 words/sec. If we assume that 18-bit words are used for all data, the transfer rate will be 1.8×10^5 bits/sec, or about 50% of the maximum for a tape. At this rate a 2400-ft tape will run for approximately 12 minutes. Depending on the parameter accuracies required and the word length used in the on-line process, we may be able to substantially reduce the data transfer rate.

Table 12.5/Example of Tape Data Rate Budget

Function	Update Rate	Word Rate
20 High-Speed Targets		
5 words each		
(r, ṙ, az, el, RCS)	20 Hz	2000 Hz
80 Slow-Speed Targets		
5 words each	5 Hz	2000 Hz
Ground Clutter Samples		
I and Q	1000 Hz	2000 Hz
Rain Clutter Samples		
I and Q	1000 Hz	2000 Hz
Headers and Overhead	1000 Hz	2000 Hz
Required Data Transfer Rate		**10,000 Hz**

In this example the data samples are updated at an average 1 kHz rate for ground and rain separately so that these may be scaled and combined as part of an on-line operator control function. Various additional data such as the clutter RCS map and antenna tables are also provided as part of a one-time initialization process.

12.6/Minicomputer Preprocessor

The simulation on-line hardware, as shown by Fig. 12.2, operates on a series of time snapshots of target data. These snapshots are stored on tape after generation by the GP computer. The functions of reading the tape in real time and the buffering to the signal generation portion of the hardware are generally accomplished by means of a minicomputer. The minicomputer provides flexibility in that its software can be modified easily. It serves as the interface between the complex set of GP computer software that generates the data and the high-speed special purpose hardware of the simulator that generates the real-time signals. Thus, in the event of interface discrepancies, the designer has the option of making modifications in a relatively small and uncomplicated software system, rather than changing the simulator hardware or regenerating the input data with the GP computer.

The operations that are logical candidates for implementing in the minicomputer can be determined primarily on the basis of their timing. There are three potential areas which must be considered. First, the minicomputer must accomplish its fundamental task, i.e., the control of the storage unit and the transfer of snapshot data from it to the data generation hardware. There is a possibility of performing some series of operations on this data during the buffering process. For example, target coordinates could be modified, the amplitudes changed, and even new simplified test targets could be generated in the minicomputer to replace some or all of those contained within the snapshot. It is also possible to modify data coming to the simulation from the radar. For example, we must decode the antenna steering signals so that they are compatible with the simulation.

The simulation must generate signals on a pulse-by-pulse basis. Search radars have pulse repetition periods that are of the order of 0.1 to 1.0 msec. This condition implies that any minicomputer operations that have to do with signal generation must be accomplished within such a time scale. Data from the storage medium, on the other hand, need be processed only a snapshot by snapshot basis. The exact time spacing between snapshots is subject to the individual requirements of a specific simulation. However, it is usually possible to provide sufficient realism with time increments that are from 10 to 100 times the pulse repetition period. Consequently, it is generally possible to undertake significant processing tasks on the buffered data within the available time slots.

Another opportunity for minicomputer processing is the interval within the radar dwell. Since the radar is grouping pulses at the dwell rate, it is often possible to slave some portion of the simulator processing to it. The dwell-to-dwell timing is usually sufficient to allow additional preprocessing of the target data by the minicomputer.

12.7/Target Signal Computation

A discrete target signal is considered to be a delayed and Doppler shifted replica of the transmitted waveform, with the amplitude being modulated by the radar two-way beam pattern. In Fig. 12.6 we illustrate the computation of the target pulse amplitude and phase based on a set of target descriptors and the radar transmission parameters. This amplitude and phase may be regarded as that resulting from a point target

response to an ideal or infinitesimally narrow pulse transmission. The target two-way range establishes the transmit-receive delay. The response to the actual transmitted pulse waveform is implemented subsequently in the output waveform generator, which we discuss in Section 12.9.

Target Descriptors

The function of the target data base generator is to provide a description of the target at each instant of time that is capable of being converted into radar signals by the interactive portion of the simulation. The description must incorporate the target properties that will affect the resultant signals. Consequently, the nature of this description is dependent upon the properties of the radar, most importantly the beam dwell time and coherent integration time. For most radar applications the following parameters present an adequate instantaneous discrete target representation:

Time
Range to radar
Azimuth of target
Elevation of target
Range rate of target to radar
Target instantaneous amplitude (or complex reflection coefficient)

If one (or more) of these target parameters does not remain constant during the radar beam dwell time or coherent integration time, then they must be varied in the midst of a beam dwell or integration period as is described in Section 2.4. While this action presents no fundamental difficulties, it can lead to complications in timing and data sequencing. Usually, the inequalities in Section 2.5 are valid so that we can assume that the above target parameters are constant during the beam dwell time or coherent integration time. The parameter that is most likely to vary is the target amplitude.

The target range, azimuth, elevation, and range rate are determined by the physical target position and motion, and as such, are not dependent on the radar parameters, except to the extent that the target may indeed be regarded as

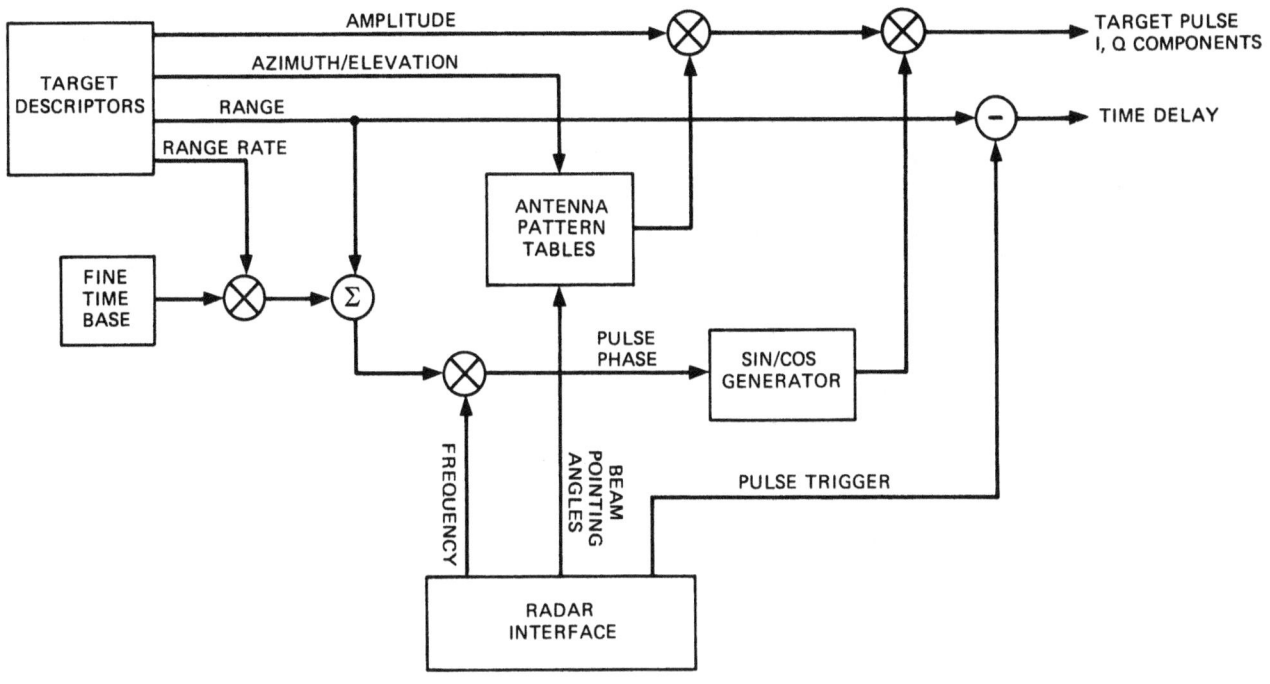

Figure 12.6/Interactive Target Signal Computation

being a point target for resolution purposes. The target amplitude, on the other hand, is a function of the radar operating band and, for large targets, of the precise operating frequency. Consequently, the target amplitude and phase descriptors may involve multiple values. If the radar can operate at several frequencies over the radar band, then each target amplitude update must include multiple RCS samples across the operating band, or suitable model parameters for use in or interpolation by the on-line portion of the simulation. For a radar that uses frequency diversity specifically for the purpose of decorrelating the target RCS fluctuations, a few random values may constitute an adequate description of the target RCS. The target amplitude descriptor includes RCS, the radar range law, and other possible scale factors such as sensitivity time control.

Antenna Beam Modulation

The target amplitude must be weighted by the two-way antenna voltage pattern. In Fig. 12.6 the azimuth and elevation antenna beam pointing angles are provided by the radar interface. These angles may vary in either a step-wise or continuous manner. It is assumed in Fig. 12.6 that the radar beam shape remains constant and that azimuth and elevation pointing data are sufficient to describe the state of the antenna beam. For a system employing a variable beam shape, the necessary control data must be provided to the simulation as the beam shape changes. In a phased array antenna the beamwidth and gain may vary as a function of scan angle off broadside. Since these are deterministic functions of the scan, the pointing angle still constitutes sufficient data for computation of the target amplitude modulation by the beam pattern. A proper choice of coordinate system can be helpful in minimizing the amount of real-time computation required, as in the case of a phased array radar where direction cosines are more natural parameters than are azimuth and elevation.

The usual example of multiple radar beamshapes is the case of a monopulse radar with independently formed sum and difference patterns. This model requires the simulation of multiple responses and the storage of multiple beamshapes. In the monopulse system the radar itself faces the problem of simultaneously processing sum and difference channel signals. Since this procedure can represent a costly replication of processing components, the radar may be configured to sequentially receive on sum and monopulse channels, or, alternately, to perform time-multiplexed processing of these signals. In such cases it is usually possible for the simulation also to take advantage of those techniques by providing sequential, rather than parallel, outputs without requiring the replication of simulation components.

The stored beam pattern tables and their subsequent use in computation present implementation questions as to the table size and number of bits that must be used throughout the on-line computations. The beam pattern quantization requirement is established by the ultimate angular accuracy of the object radar, which can result in the requirement for a small angle and amplitude quantization step size for a precision tracking radar. The angle quantization rms error is $1/\sqrt{12}$ of the angle step size, or

$$\mathcal{E}_\theta = 0.29\,\Delta\theta \qquad (12.6)$$

where $\Delta\theta$ is the angle step size, assuming a random distribution of radar beam pointing angles around the target and no interpolation. A beam pattern table size problem quickly develops. If, for example, a simulation rms accuracy of .01 beamwidth and a simulated pattern covering ± 2 beamwidths are desired in azimuth and elevation, then a table size of over 13,000 entries is

required for the sum pattern alone. This data must be provided in a fast access memory since it is used in a (near) real-time interactive operation.

Usually, only the antenna pattern in the vicinity of the mainbeam is of interest for simulating targets, as discussed in Section 6.4. There are exceptions, though, such as the multipath return that appears in the antenna sidelobe region; sometimes, extremely intense false targets such as aircraft are of interest when they appear in sidelobes. In those cases we have to include some description of the antenna pattern that goes farther away from the mainlobe than just a few beamwidths. However, such a description can be rather coarse.

In many cases it is possible to take advantage of the particular antenna pattern characteristics to greatly reduce the antenna table size as in the case of a rectangular antenna using independent azimuth and elevation aperture illumination functions. The combined pattern may be expressed as a product of two one-dimensional functions. For some circular or elliptical antennas the pattern may be described as a one-dimension function of a simply computed function of azimuth and elevation. The antenna table size requirement may also be lessened somewhat by the use of a variable angular step size.

Finally, that size requirement may be greatly reduced (at the expense of some additional computation) by interpolating between pattern samples, as discussed in Section 12.3. In the above example linear interpolation provides over a 50:1 table size reduction while maintaining an equivalent angle error of .01 beamwidths. However, the use of a moderate table size without interpolation generally is preferable if there is some acceptable antenna pattern approximation. The number of bits to be used in the antenna tables and subsequent computations also is established by the desired simulation accuracy. The rms angular error, σ_θ, resulting from the use of n bits in the quantization of the antenna patterns is given, near boresight, by

$$\sigma_\theta = \frac{\theta_{3dB}}{\sqrt{3}\,k\,2^n} \tag{12.7}$$

where θ_{3dB} is the half-power beamwidth and k is the error slope which is usually of the order of 1.5 for a monopulse system. This expression is based on the generalized error expressions given by Barton [Ref. 9]. In Eq. 12.7, θ_{3dB} is the half-power width of the sum pattern. If a .01 beamwidth rms simulation accuracy is desired, for example, then 6 bits of significance are required for the antenna table values. Off-boresight, the errors are magnified so that somewhat better quantization may be required. A second restriction on the antenna table word size is that there be sufficient dynamic range to accommodate the sidelobe level at which large false targets may be detectable. Since this dynamic range may exceed 50 dB, a larger number of bits may be required.

Target Phase Computation

The computation of the received phase involves the target range, range rate, instantaneous radar frequency, and the phase associated with the target fluctuation; we regard the target as a discrete point in space characterized by an amplitude and phase, and possibly a function of the transmitted frequency. The instantaneous received phase ϕ is given by

$$\phi = 4\pi(r + \dot{r}t)\lambda + \phi_0 \tag{12.8}$$

where r, $\dot{r}$, and ϕ_0 are the range, range rate, and fluctuation phase of the target, respectively, and λ is the instantaneous transmission wavelength. The time, t, is measured from the time of the last update of the target range.

The computation of the phase in Eq. (12.8) can be expensive in real time if the range must be specified to within a fraction of a wavelength, requiring over 24 bits of data computation in typical cases. Fortunately, as is described in Section 5.1, it is adequate to treat the absolute phase associated with the target reflection properties as being random. However, the *change* in

relative phase (as a function of frequency between two targets within the same range resolution cell) should be preserved, but many fewer bits are now required for range in Eq. (12.8). Finally, the absolute phase, ϕ_0, can usually be ignored altogether without any loss in generality. The roundoff errors associated with the quantization of r provide a degree of pseudo-randomness in the phase computation.

The number of bits necessary to describe phase is straightforward to determine. A quantization of $\Delta\phi$ radians in the phase computation results in the generation of spurious Doppler signals, the rms magnitude of which is given by

$$\varepsilon_A = \frac{A \cdot \Delta\phi}{\sqrt{12}} \tag{12.9}$$

where A is the signal amplitude. If we quantize phase with n bits, then $\Delta\phi = 2\pi/2^n$ and

$$\varepsilon_A = 1.81 \times 2^{-n}\,A \tag{12.10}$$

The general rule is to choose n large enough so that, for the maximum value of A, the resulting value of ε_A is lower than the receiver noise magnitude. For n = 10 bits, we can accommodate a dynamic range of 55 dB. However, in anticipation of additional errors that follow, we should allow a margin of at least 6 dB.

It is usually necessary to convert the computed amplitude and phase to the signal in-phase and quadrature components. If we use the table lookup methods in Section 9.3 for the sine/cosine calculation, the input phase is quantized automatically. The number of table entries over the first quadrant is given by 2^{n-2} where n is the above number of bits. However, another problem arises concerning the number of bits required for the signal components. If q is the quantization step size, then the rms phase error for a phasor at the cardinal points ($\phi = 0$, $90°$, $180°$, $270°$), assuming no other errors, is given by

$$\varepsilon_\phi = \frac{1}{\sqrt{12}}\,\frac{q}{A} \tag{12.11}$$

where A is the phasor magnitude. As the phasor rotates, the rms error increases to a maximum of $\sqrt{2}$ times Eq. (12.11). Using this new value, we can write the rms magnitude of the spurious Doppler signals as

$$\mathcal{E}_A = \mathcal{E}_\phi A \simeq 0.41\, q \qquad (12.12)$$

which is independent of A. We should also choose q so that this value of $\mathcal{E}_A$ will be less than the receiver noise magnitude (q equal to the noise magnitude would be adequate). The number of bits, n, required for each phasor component must satisfy $2^{n-1} \geqslant q/A_{max}$, where A_{max} is the maximum signal amplitude. We should choose n not only for the signal phasor components, but also for the entries in the sine/cosine table. On the other hand, since we have to accommodate the dynamic range of only the target signals, we might be able to get by with fewer bits to describe the signal amplitude prior to computing the phasor components.

12.8/Distributed Clutter Generation

The generation of radar clutter signals presents simulation design problems of a different character than those encountered in the target signal case. Targets essentially are defined by their discrete, low-density nature and their possession of a trajectory in real space. Distributed ground and rain clutter, on the other hand, is characterized by its continuous high-density nature, i.e., a high percentage of contiguous radar range-azimuth resolution cells contain clutter signals. However, clutter is not normally considered to possess a trajectory in the sense that a discrete target does. The prime properties of continuous clutter that must be incorporated into a simulation are its spatial distribution and fluctuation characteristics.

Clutter generation has its greatest impact on the on-line portion of the simulation; the simulation output data volume requirements for clutter are too great for off-line pregeneration and direct transfer to the on-line components. Continuous clutter returns from the environment, such as those from ground and rain, may be broken down into elemental returns in each radar resolution cell. Each range-azimuth-elevation resolution cell observes returns from physically different volume elements of space. Resolution cell returns may be related through their underlying statistics and models, but the actual histories are statistically independent from cell to cell because different sets of scatterers are involved in each cell.

A radar receives independent clutter environment samples at a rate approximately equal to twice the bandwidth of the transmitted pulse. The simulation need only generate clutter signals in those areas of space which are cluttered, which may be as low as 10% of the total radar processing volume; nonetheless, for megahertz-bandwidth radars the average data output rate required by the simulation is still hundreds of kilohertz. This rate cannot be accommodated by first pregenerating the data and then transferring it in real time because the data volume becomes prohibitively large. Nor can we compute it entirely in real time since the computation rate would be unreasonable.

The solution to this problem, therefore, is in the use of an on-line clutter sample generation technique that is based on a limited pregenerated data base describing the clutter. Because of the high output data rate requirement, the amount of on-line computation to be performed for each sample must remain as small as possible to minimize costly computation hardware.

Clutter Random Process Generation

The most important attributes of clutter to be reflected in the output signals of the simulation are spatial nonhomogeneity, statistical fluctuation, and spectral shape. Of these, the last poses the most difficult problem for on-line signal generation, due to the lack (in all but special cases) of simple computational techniques for generating random processes that have specific spectral shapes. The special cases correspond to the processing of white noise by a linear lumped parameter analog network, a recursive digital filter, or a weighted tapped-delay line. Although such processors can be designed to generate a variety of possible spectral shapes, they unfortunately do not produce those of greatest interest for clutter spectral modeling, i.e., the Gaussian spectrum, $1/f^3$ spectrum, and clutter spectra possessing an antenna related sidelobe structure. For some applications an approximation to these desired shapes is acceptable; but unfortunately the approximate spectral shapes achievable from networks such as the above usually diverge from the desired shapes most predominantly on the spectral tails — the region of greatest interest in signal simulation for MTI radars. More suitable approaches for generating clutter processes with the desired spectral shapes are described in Chapter 8.

Since each radar spatial resolution cell observes returns from physically different areas of clutter, the radar signal simulation ideally has an independent clutter generator for each such resolution cell. Approaches to approximating this condition will now be discussed.

Generating Clutter Signal in a Resolution Cell

The clutter signal appearing in one resolution cell can be assumed to be a stationary random process over the beam dwell time, characterized by an average power, a spectral shape, and a mean Doppler frequency. In Chapter 8 there are several methods discussed for generating correlated time sequences that have those properties.

However, real-time computation was not then a consideration.

In Table 12.6 we list three possibilities for generating the clutter signals. In general, the two extremes, generating the clutter signals either entirely off- or on-line, are not practical. In the former case, the data transfer rate is almost always too high; in the latter, the computational burden is excessive. As a compromise, it is advantageous to generate the clutter data at a lower sample rate off-line and then, on-line, interpolate and resample the data at the PRF. Notice that in a non-real time simulation clutter samples can be generated at the PRF and no interpolation is required, but that in the real-time case the radar PRF is usually not known *a priori*. In the process of resampling the reconstituted data, the interpolation spectral residue can fall anywhere within the PRF unambiguous frequency zone whenever the PRF is not equal to the original sampling rate of the data. The simulation generated clutter spectrum as seen by the radar is depicted in Fig. 8.9.

As is discussed in Section 8.5, there is a trade-off between the number of points used in the interpolation and the sampling rate. Sample holding is the simplest method to implement in real-time, but it requires the highest sampling rate. Linear, or even quadratic interpolation, is more complex to implement, but the sampling rates to achieve the same overall simulation accuracy can be much lower. To illustrate this trade-off, let us assume that the half-power spectral width of the clutter is $f_{3dB} = 60$ Hz and we desire to keep the spectral residue due to sampling below about -60 dB. From Eq. (8.61) or from Fig. 8.10 we can derive the minimum sampling rate prior to interpolation as

f'_r = 25,000 Hz (hold data)

    = 1,000 Hz (linear interpolation)

    =   400 Hz (quadratic interpolation)

The minimum sampling rate of 25,000 Hz (complex) with no interpolation is not attractive since

Table 12.6/Approaches to Generating Clutter Samples

Off-Line Computation	On-Line Computation	Advantage	Disadvantage
Generate random sequence at high sample rate	scale	minimal real-time computation	high data transfer rate
Generate random sequence at lower sample rate	interpolate, resample, and scale	low data transfer rate	significant real-time computation
none	Generate random sequence at PRF	minimal data transfer	excessive real-time computation

it almost equals the tape maximum data transfer rate. However, the rate of 1,000 Hz with linear interpolation is appealing. The added computational burden associated with quadratic interpolation is an overriding disadvantage since the rate can be reduced only by a factor of 2.5 compared to linear interpolation.

Multiple Clutter Process Generation

The received radar signal for clutter consists of independent processes within each radar resolution cell. If we are simulating N resolution cells, there will be N independent clutter signals. However, the received signal from each element of the same type of clutter can be modeled as belonging to the same underlying random process over large regions of space, where only different amplitudes are applied to each cell corresponding to some spatial nonhomogeneity. The clutter spectrum as seen by a moving radar platform can be expected to vary more rapidly with both antenna scan angle and range, but nonetheless, it remains approximately constant over a large number of otherwise independent resolution cells.

It is possible in the real-time simulation for us to take advantage of the consistency of the clutter spectrum by making use of a single clutter random process having the desired spectrum to generate returns for many pseudo-independent resolution cells. Although there may be N resolution cells being simulated, we are able to simulate fewer than N independent clutter signals.

Thus only a few random process generators having a few different desired spectral and statistical properties are able to serve as an adequate underlying data base for all real-time clutter generation. The disadvantages of this approach are minor compared to the otherwise overwhelming problems of data transfer or true on-line signal generation.

In Fig. 12.7 we illustrate a method for generating multiple pseudo-independent clutter processes from a single input process. The approach relies on the fact that linear combinations of independent random processes having the same spectrum also have that spectrum. The delay line tap separations that provide M signals designated by $y_m(t)$ in Fig. 12.7 are separated by nonuniform delays, τ_m, which are a few times greater than the correlation time, τ_c, of the generating process, and thus represent approximately independent segments of the signal over the time span of the smallest value of τ_m. Linear combinations of the signals, $y_m(t)$, with scalar coefficients, a_{mn}, produce multiple processes, x_n, as

$$x_n(t) = \sum_{m=1}^{M} a_{mn} y_m(t), \quad n=1,\ldots,N \qquad (12.13)$$

The signals, $x_n(t)$, have the same spectrum, but are partially correlated. Since the same samples progress down the delay line, there are delayed time replications given by

$$y_\ell(t) = y_m(t + \tau_\ell - \tau_m) \qquad (12.14)$$

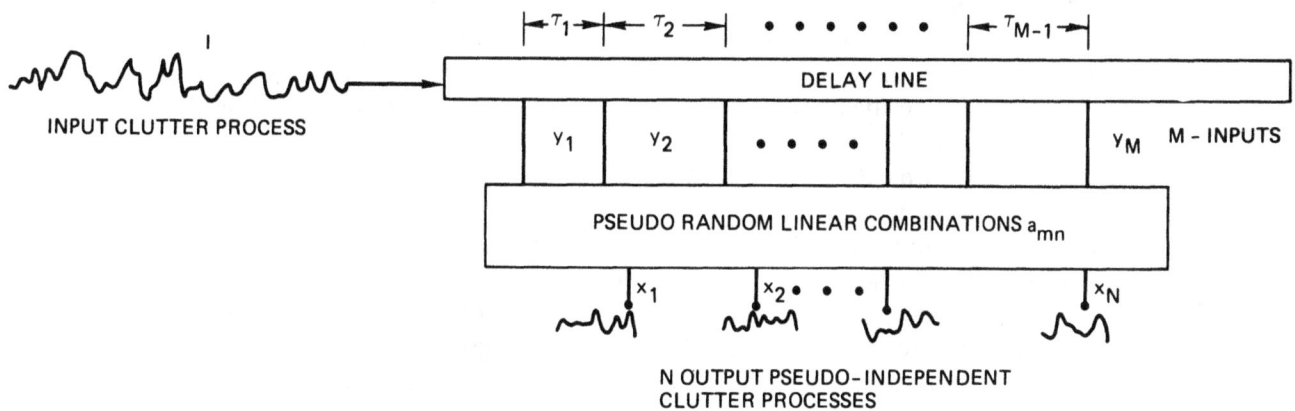

Figure 12.7/Generation of Multiple Pseudo-Independent Processes

If the number of correlation intervals, M, stored on the line in Fig. 12.7 is too small, there is a distortion of the low frequency content of the resulting signal spectrum of $x_n(t)$.

Let us define the correlation coefficient between two different signals as

$$\rho_{kn}(\tau) = \overline{x_k(t)x_n^*(t - \tau)} \Big/ \overline{|x_n|^2} \qquad (12.15)$$

where the bar designates an ensemble average. (The random processes are stationary; the average could also be over time.) The optimum values of k, n, and τ that would give consistently low values of Eq. (12.15) are not known. But if τ_m and a_m are chosen in a pseudo-random manner with $\overline{\tau_m} > \tau_c$ and $\overline{a_{mn}^2} = 1$, then the average correlation coefficient magnitude can be shown to approach

$$\text{AVG}\ [|\rho_{kn}(\tau)|]\ \rightarrow \sqrt{2/\pi M} \qquad (12.16)$$

where the average is over all k and n. Thus, for M of the order of a few hundred, low correlation coefficients can be obtained.

The implementation of multiple clutter processes with low correlation via the method in Fig. 12.7 still involves a significant amount of on-line computation. This operation may be greatly reduced by selecting $a_{mn} = 0$ except for a subset of J values of m = 1, . . . , M, for which $a_{mn} = \pm 1$. When a_{mn} are chosen in this way the average correlation between any two $x_n(t)$ remains bounded by Eq. (12.16), but the correlation coefficient distribution becomes less homogeneous. It is difficult to specify a value for J without taking the simulation objectives into consideration. A small value of J decreases the number of computations required at the expense of a more cyclic behavior in the clutter processes. Reasonable choices for J lie in the range from 3 to 20. Regardless of the size of J the number of pseudo-independent processes that can be generated in this way is several times M.

The significance of a non-zero correlation coefficient between the clutter samples depends upon the effects of clutter on the radar with which the simulation is used. The two most common areas in which clutter statistics have an impact are in the false alarm rate and in adaptive threshold processing [Ref. 10]. Both involve the averaging over large numbers of processed clutter samples; the existence of some clutter correlation does not significantly affect the results of either the false alarm or adaptive threshold statistics.

By the use of the above techniques it is possible to generate random processes in several hundred or several thousand pseudo-independent clutter resolution cells, all having the same underlying clutter spectrum, and all created from a single random process which is generated off-line and passed to the on-line portion of the simulation. The data transfer update rate is thereby proportionately reduced. Depending on the application, this rate can be further reduced by using the discrete Fourier transform, as in Eq. (8.22), to generate the original random sequence. Such a sequence has the property of being circularly repetitive in the sense that the first sample corresponds to the one that would otherwise follow the last. The random sequence recirculates. This approach is often acceptable in cases where the radar has a short processing memory. As an example, we can choose the following values:

Clutter spectral width	f_{3dB}	= 60 Hz
Sample generation rate	f_r'	= 1,000 Hz
Number of clutter samples generated by DFT	N_r	= 4096
Spectral sampling resolution	Δf	= 0.244 Hz
Number of clutter correlation intervals	M	= 246
Number of clutter sequences summed	J	= 5

With these values it is possible to generate several hundred pseudo-independent clutter processes with average cross-correlation of approximately 5%. For linear interpolation on the $x_n(t)$ samples the clutter spectral residue is down by 59 dB.

Spatial Clutter Map
The next most important clutter property in the simulation is its spatial nonhomogeneity, which must be applied to the above generated resolution cell samples. Depending upon the radar parameters and the terrain of interest, these nonhomogeneities may encompass from a few radar clutter resolution cells to several hundred. The clutter map should be in a radar space coordinate system to minimize the real-time computa-

tion, and it should include the weighting by the antenna pattern, including integration over sidelobes. In most ground based radar applications it is reasonable to reduce the elevation dimension of the map to a separate elevation weighting variable. The size of the map that must be stored is a function of the roughness of the spatial clutter variations, the antenna beamwidths, and whether or not interpolation is performed on the map outputs.

Noise Generation
If the simulation output is to be injected into the radar receiver at video, then the simulation must provide noise signals. Pseudo-random independent noise samples may be generated by any of several means including those discussed in Section 9.2 and by analog thermal noise generation. In many cases it is adequate to generate uniformly distributed pseudo-random digital noise. If it is produced at a rate many times higher than the radar IF or video bandwidth, then those linear filtering functions have the effect of converting the uniformly distributed noise into that which is approximately Gaussian distributed. When the simulation output signals are injected after the radar sensitivity time control function, the range scaling of the noise magnitude must be applied by the simulation.

12.9/Output Waveform Generator

The output waveform generator is the final stage of the real-time simulation. It converts the target and clutter computed samples into true real-time outputs through synchronization with the transmitted waveform. The output signal D/A conversion and the transformation to the radar

IF or RF frequency are also performed by this function. If the output signal is to be injected into the radar at video as in-phase and quadrature components, then the additional filtering of the by-passed IF filters must be provided by the simulation output interface unit with equivalent video filtering.

Target, clutter, and noise inputs to the output waveform generator are computed by the target and clutter generators for each radar pulse. These in-phase and quadrature signals are the discrete target and clutter responses to an ideal or infinitesimally narrow transmitted pulse. The output waveform generator time-sequences these outputs according to their delay after the transmitted pulse trigger, as shown in Fig. 12.8. An additional time delay must also be accounted for that corresponds to the system delay forward of the point at which simulated signals are injected. The so-generated target, clutter, and noise samples are summed to obtain the composite ideal pulse radar return signal.

The simulated return from the true radar transmitted pulse waveform with its finite resolution properties is obtained by convolving the ideal infinitesimally narrow pulse response with the actual transmitted pulse waveform. If the radar transmitted waveform is a simple pulse, then this operation, combined with the video level equivalent IF filter, is a simple filtering which can be accomplished with analog components after D/A conversion. If the transmitted pulse is coded with a relatively low time-bandwidth product, then the convolution operation can be implemented by means of a tapped-delay line, possibly with complex coefficients, followed by suitable analog filtering. Unfortunately, for other classes of pulse waveforms, the convolution operation can become quite cumbersome since the convolution must be performed with special purpose high-speed hardware.

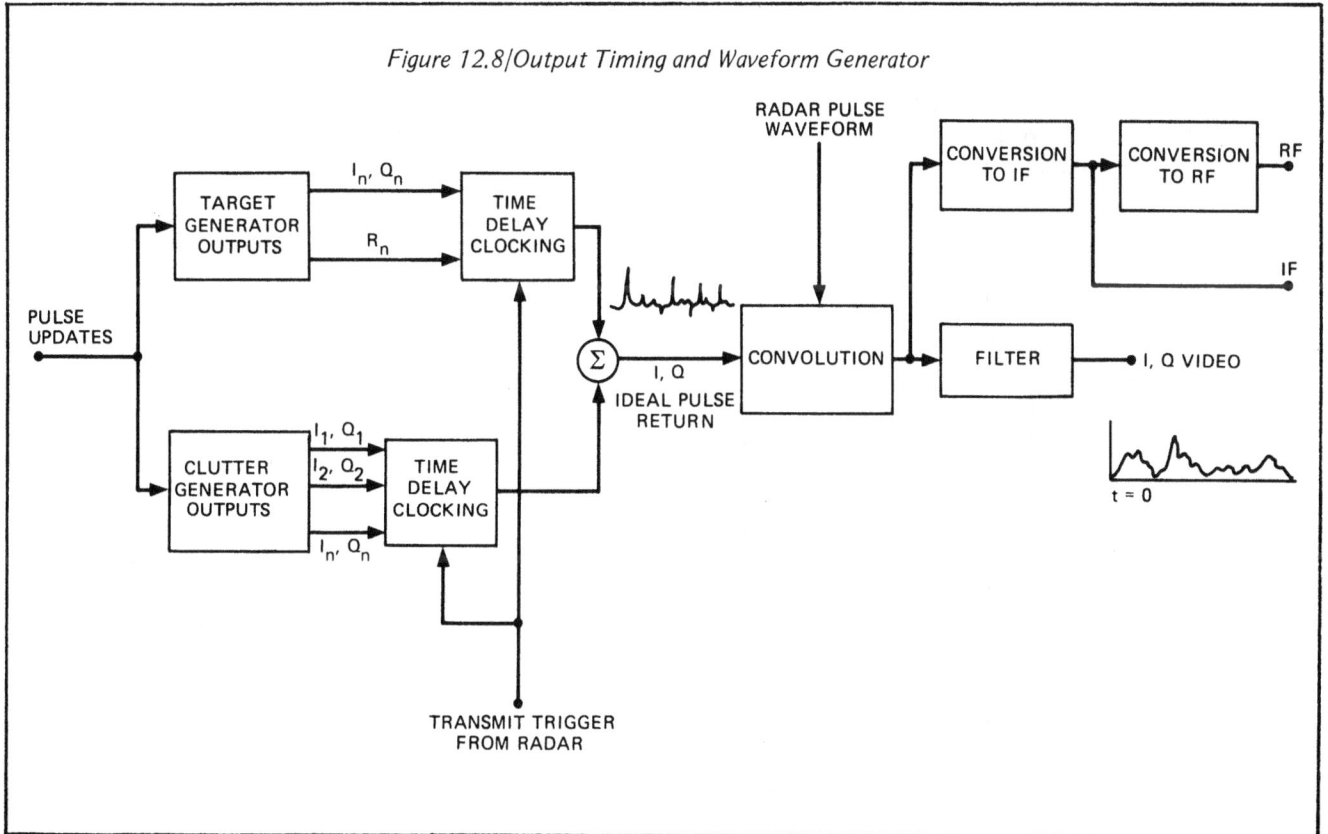

Figure 12.8/Output Timing and Waveform Generator

The waveform response convolution operation and associated filtering provide the actual response of the radar pulse waveform to the simulated environment. In the implementation of this step, we assume that there is no Doppler resolution on the basis of a single pulse. If this were not the case then a one-dimensional convolution operation would not produce an accurate representation of the received signal. To accurately treat such a situation would require an extreme complication, because each pulse would have to be generated individually for each target. When this case occurs it may be more reasonable to inject the simulated signal into the radar signal stream after pulse compression has been performed, since the presumed primary objective of the simulation is to exercise the subsequent signal and data processing, rather than to examine the waveform ambiguity function properties.

The in-phase and quadrature component signals at the output of the pulse waveform convolution operation represent the final simulated video level signals. They are suitable for up-conversion to the radar IF or RF frequencies, using reference signals supplied by the radar, for injection at suitable points in the receiver. However, it is prefereable, whenever possible, to use the simulated signals directly at the video level — the point where the greatest signal fidelity and dynamic range are achieved. Although the simulated signal dynamic range typically exceeds 50 dB, maintaining this range and signal fidelity in the process of vector modulation to the IF or RF frequency can be an exceedingly difficult, if not impossible, task. Furthermore, little additional value is lent to the simulation through signal injection at IF or RF, since the primary value of an approach that maintains a high degree of signal fidelity is in the exercising and testing of the video signal processing, data processing, and other downstream operations. If the simulated signals are to be inserted into the radar at the baseband video level, then radar processing operations preceeding this point must be accounted for or mimicked in the simulation process. Such operations may include IF filtering, limiting, STC, and pulse compression.

12.10/An Example

A discussion of principles and approaches to real-time radar environment signal simulation makes little sense unless such a simulation is practically and economically feasible. In the foregoing sections the emphasis has been on underlying considerations with example approaches being outlined for many of the key simulation functions. In many cases it has been necessary to rely on stated or implied characteristics of the object system, usually a ground based pulsed radar, to avoid inevitable impracticalities of overgeneralized approaches. In this section we describe an example of an operational real-time radar signal simulation that makes use of many of the principles and approaches outlined in the previous sections. This simulation is the Radar Environment Simulator (RES), which has been developed by Technology Service Corporation* for a class of ground based high-performance phased array radars. The characteristics of these radars cover a wide range, but all radars are serviced by similar on-line simulation units having different off-line generated data bases. The radars include mechanical, phase, and frequency one-dimension scan, and phase-phase and phase-frequency two-dimensional scan antenna types. Beamwidths vary from 1° to 8° and scan sectors up to 90°. The radar PRF's and bandwidths vary from approximately 2 kHz to 20 kHz and from .5 MHz to 10 MHz, respectively. Virtually all

*The program has been sponsored by the U.S. Army Electronics Command, Fort Monmouth, New Jersey. At the time of this writing, three such real-time simulation units have been built and are in use with different operating radars.

radar bands are represented. Signal processing types include digital pulse compression, digital MTI cancelers, and Doppler filter banks. Power levels vary from a few hundred watts to several kilowatts. Both monopulse and sequential lobing tracking are represented. The radars also employ computer processing and control and are highly interactive with the environment. These radar characteristics do not represent limits on a practical real-time simulation, but illustrate the range of radar characteristics with which such a simulation has been successfully operated.

To date over 10,000 hours of simulation operating time have been accumulated in use with the radars. This operating time has covered all phases of radar development, (including hardware and software debugging, performance evaluation, operator training, status checking and use of the simulation as the driving element in overall system exercises.) The total cost of the real-time simulations and their operation has been estimated at less than ten percent that of their live test mission equivalent. As a development test

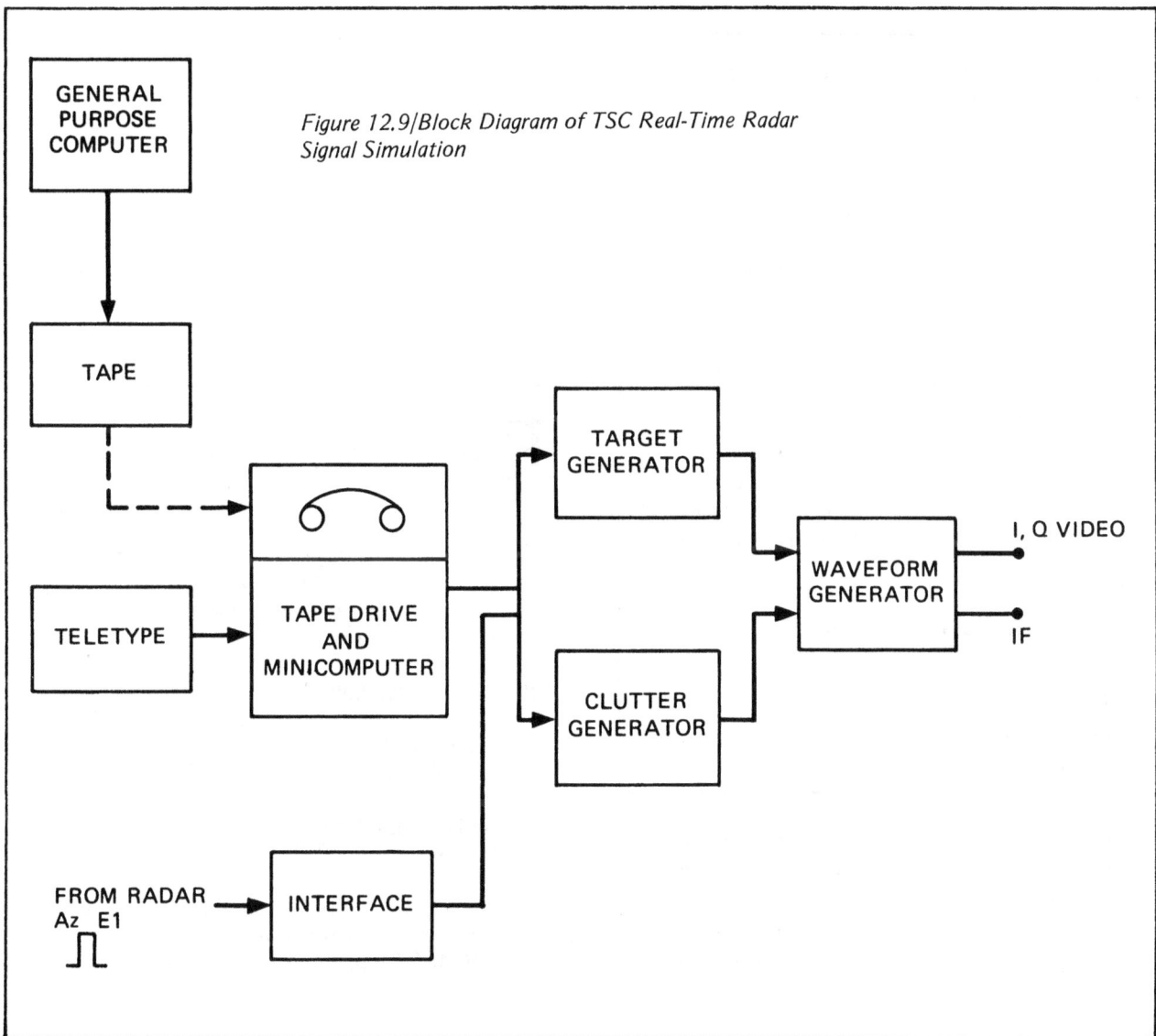

Figure 12.9/Block Diagram of TSC Real-Time Radar Signal Simulation

bed, the simulations have been credited with significant savings in the radar development time.

In Fig. 12.9 we show a block diagram, and in Fig. 12.10 a photograph, of the TSC Radar Environment Simulator. The right-hand rack houses the magnetic tape transport unit and the minicomputer. The left-hand electronics rack houses the three principle functions of target generation, clutter generation, and the output waveform synthesis. These simulations operate essentially according to the principles and functional block diagrams discussed in this chapter.

Table 12.7. The on-line minicomputer has been used in the Radar Environment Simulator to provide a high degree of operator on-line control. Through teletype-entered commands during initalization, the operator selects ground and rain clutter spatial distributions, spectral shape, bandwidth, and velocity offset from several models that have been prestored on tape. The simulation target scenario can be configured on-line as desired from those target flights which have been provided in the pregenerated tape data base.

Table 12.7/Typical Characteristics of TSC Real-Time Radar Environment Simulator

TARGETS

Number of Simultaneous Targets	30 High speed plus 100 Slow speed
Parameter Quantization	.12 Milliradian azimuth/ elevation 15 Meters range .40 dB RCS .90 m/s Velocity
Parameter Range	± 60° azimuth ± 30° elevation 270 km range ± 1000 m/s velocity 48 dB dynamic range

CLUTTER

Number of Range Gates Per Beam	128
Dynamic Range	54 dB
Spurious Signals	< -54 dB

In the implementation of the on-line real-time simulation it was necessary to impose some limitations on the number of targets, accuracy, and amount of clutter that could be generated. The resulting typical characteristics are indicated in

References

[1]   Greenstein, L.J, A.E. Brindley, and R.D. Carlson; "A Comprehensive Ground Clutter Model for Airborne Radars," *IIT Research Institute (Chicago, Illinois) Report to Overland Radar Technology Program*, September 15, 1969.

[2]   Fishbein, W., S.W. Graveline, and O.E. Rittenbach; "Clutter Attenuation Analysis," *USAECOM Technical Report ECOM-2808*; Ft. Monmouth, N.J., March 1967.

[3]   Nathanson, F.E.; *Radar Design Principles*; McGraw-Hill, New York, 1969.

[4]   Pollon, G.E.; "Distribution of Radar Angels," *IEEE Trans. Aerospace and Electronic Systems*, Vol. AES-8, pp. 721-727, November 1972.

[5]   Eastwood, E.; *Radar Ornithology*; Methuen & Co., London, 1967.

[6]   Konrad, T.G., J.J. Hicks, and E.B. Dobson; "Radar Characteristics of Birds in Flight," *Science, Vol. 159*, pp. 274-280, January 19, 1968.

[7]   Hardy, K.R. and I. Katz; "Probing the Clear Atmosphere with High Power, High Resolution Radars," *Proc. IEEE, Vol. 57*, pp. 468-480, April 1969.

[8]   Skolnik, M.I. (ed.); *Radar Handbook*; McGraw-Hill, New York, 1970.

[9]   Barton, D.K., and H.R. Ward; *Handbook of Radar Measurements*; Prentice-Hall, 1969.

[10] Finn, H.M., and R.S. Johnson; "Efficient Sequential Detection in the Presence of Strong Localized Signal Interference," *RCA Review*, September 1966.

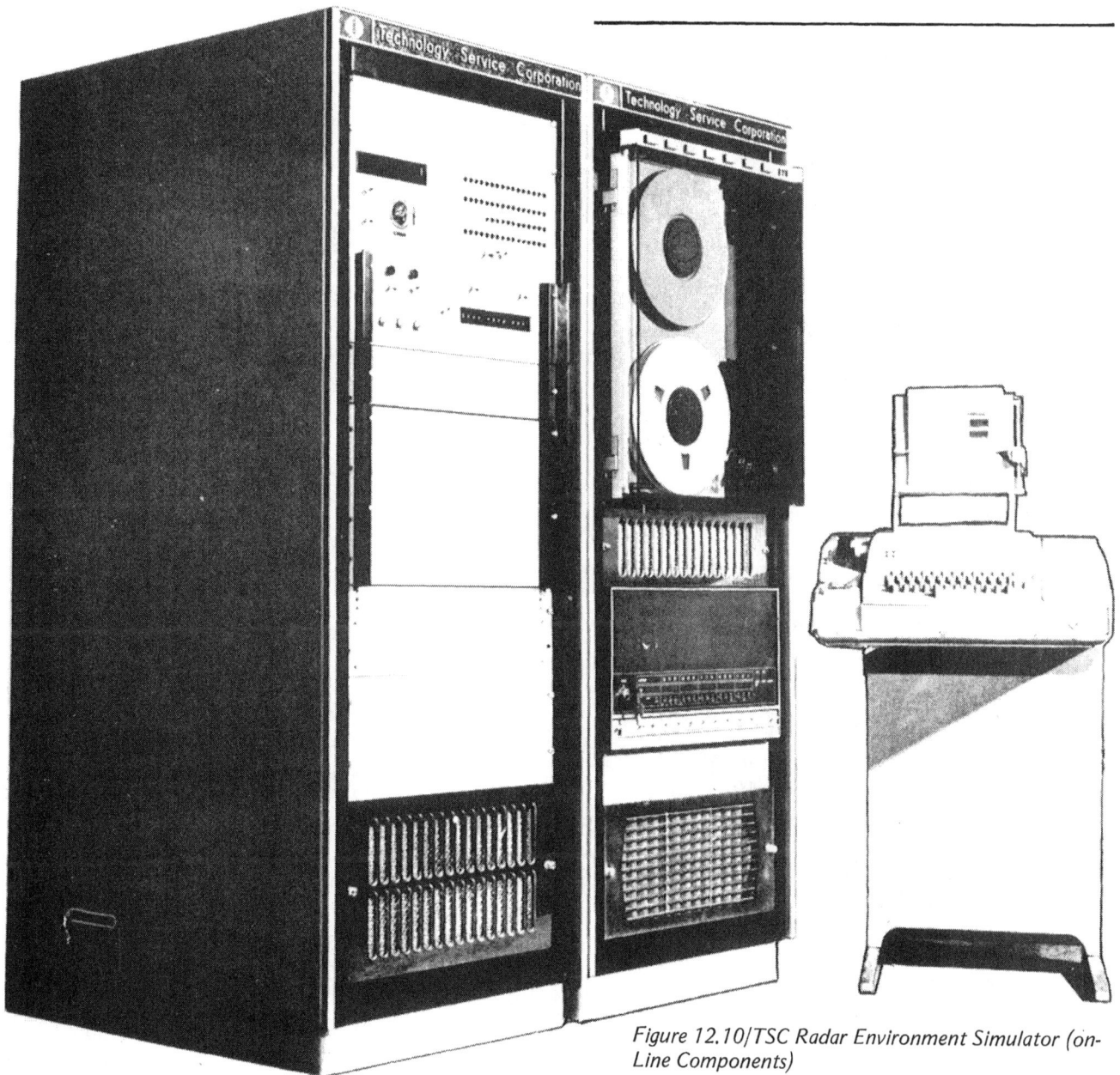

Figure 12.10/TSC Radar Environment Simulator (on-Line Components)
Photo courtesy of Technology Service Corporation.

Index